Understanding Biodiversity

Understanding Biodiversity

Arianna Reeves

R CALLISTO REFERENCE

www.callistoreference.com

Callisto Reference,
118-35 Queens Blvd., Suite 400,
Forest Hills, NY 11375, USA

Visit us on the World Wide Web at:
www.callistoreference.com

ISBN: 978-1-64116-598-3 (Hardback)

Cataloging-in-Publication Data

Understanding biodiversity / Arianna Reeves.
   p. cm.
Includes bibliographical references and index.
ISBN 978-1-64116-598-3
1. Biodiversity. 2. Biology. 3. Biocomplexity. 4. Numbers of species.
5. Ecological heterogeneity. I. Reeves, Arianna.

QH541.15.B56 U53 2022
333.95--dc23

TABLE OF CONTENTS

PREFACE

This book is a culmination of my many years of practice in this field. I attribute the success of this book to my support group. I would like to thank my parents who have showered me with unconditional love and support and my peers and professors for their constant guidance.

The variability of life on earth is known as biodiversity. It is primarily concerned with the measurement of the variation at the genetic, species and ecosystem level. Important sub-fields of biodiversity include terrestrial biodiversity and marine biodiversity. Terrestrial biodiversity is found in the areas near the equator due to warm climate and high primary productivity. The diversity of all living beings is based on their habitat and factors such as temperature, altitude soils, geography, soils, and precipitation. The field known as biogeography is also related to the biodiversity. It is the study of the spatial distribution of organisms, species and ecosystems. This book is a compilation of chapters that discuss the most vital concepts in the field of biodiversity. Such selected concepts that redefine this field have been presented in this book. The topics covered in it offer the readers new insights in this field.

The details of chapters are provided below for a progressive learning:

Chapter – Introduction

The measure of variation at the level of species, genetics and ecosystem of the Earth is known as biodiversity. The main types of biodiversity include genetic diversity, species diversity and ecosystem diversity. This chapter will provide a brief introduction to the different aspects of biodiversity and its various types.

Chapter – Evolution: Types and Processes

The change in heritable characteristics of the biological populations is referred to as evolution. It is broadly divided into a few categories such as divergent evolution, convergent evolution, macroevolution and microevolution. All these diverse processes and types of evolution have been carefully analyzed in this chapter.

Chapter – Threats to Biodiversity

Factors such as pollution, overexploitation and climate change can cause loss of biodiversity. The species which are adversely affected by these factors are categorized as threatened, vulnerable, endangered, critically endangered and extinct in the wild. These causes of biodiversity loss as well as the different classifications of threatened species have been thoroughly discussed in this chapter.

Chapter – Biodiversity Conservation

The management of natural resources with an aim to sustain biodiversity in ecosystems, species and the evolutionary processes is known as conservation of biodiversity. This chapter has been carefully written to provide an easy understanding of the importance of biodiversity and the diverse types of conservation such as ex situ conservation and in situ conservation.

Chapter – Five Kingdom Classification

The organisms have been classified into five kingdoms on the basis of likeness and variances among them. Monera, protist, fungus, plant and animal are the five kingdoms of organisms. The topics elaborated in this chapter will help in gaining a better perspective about these classifications of organisms.

Arianna Reeves

Introduction

The measure of variation at the level of species, genetics and ecosystem of the Earth is known as biodiversity. The main types of biodiversity include genetic diversity, species diversity and ecosystem diversity. This chapter will provide a brief introduction to the different aspects of biodiversity and its various types.

BIODIVERSITY

- Biodiversity is the variability among living organisms from all sources, including terrestrial, marine, and other aquatic ecosystems and the ecological complexes of which they are part; this includes diversity within species, between species, and of ecosystems.

- Biodiversity forms the foundation of the vast array of ecosystem services that critically contribute to human well-being.

- Biodiversity is important in human-managed as well as natural ecosystems.

- Decisions humans make that influence biodiversity affect the well-being of themselves and others.

Biodiversity is the foundation of ecosystem services to which human well-being is intimately linked. No feature of Earth is more complex, dynamic, and varied than the layer of living organisms that occupy its surfaces and its seas, and no feature is experiencing more dramatic change at the hands of humans than this extraordinary, singularly unique feature of Earth. This layer of living organisms—the biosphere—through the collective metabolic activities of its innumerable plants, animals, and microbes physically and chemically unites the atmosphere, geosphere, and hydrosphere into one environmental system within which millions of species, including humans, have thrived. Breathable air, potable water, fertile soils, productive lands, bountiful seas, the equitable climate of Earth's recent history, and other ecosystem services are manifestations of the workings of life. It follows that large-scale human influences over this biota have tremendous impacts on human well-being. It also follows that the nature of these impacts, good or bad, is within the power of humans to influence.

Biodiversity includes all ecosystems—managed or unmanaged. Sometimes biodiversity is presumed to be a relevant feature of only unmanaged ecosystems, such as wildlands, nature preserves, or national parks. This is incorrect. Managed systems—be they plantations, farms, croplands, aquaculture sites, rangelands, or even urban parks and urban ecosystems—have their own biodiversity.

Given that cultivated systems alone now account for more than 24% of Earth's terrestrial surface, it is critical that any decision concerning biodiversity or ecosystem services address the maintenance of biodiversity in these largely anthropogenic systems.

Measuring Biodiversity: Species Richness and Indicators

In spite of many tools and data sources, biodiversity remains difficult to quantify precisely. But precise answers are seldom needed to devise an effective understanding of where biodiversity is, how it is changing over space and time, the drivers responsible for such change, the consequences of such change for ecosystem services and human well-being, and the response options available. Ideally, to assess the conditions and trends of biodiversity either globally or sub-globally, it is necessary to measure the abundance of all organisms over space and time, using taxonomy (such as the number of species), functional traits (for example, the ecological type such as nitrogen-fixing plants like legumes versus non-nitrogen-fixing plants), and the interactions among species that affect their dynamics and function (predation, parasitism, competition, and facilitation such as pollination, for instance, and how strongly such interactions affect ecosystems). Even more important would be to estimate turnover of biodiversity, not just point estimates in space or time. Currently, it is not possible to do this with much accuracy because the data are lacking. Even for the taxonomic component of biodiversity, where information is the best, considerable uncertainty remains about the true extent and changes in taxonomic diversity.

There are many measures of biodiversity; species richness (the number of species in a given area) represents a single but important metric that is valuable as the common currency of the diversity of life—but it must be integrated with other metrics to fully capture biodiversity. Because the multidimensionality of biodiversity poses formidable challenges to its measurement, a variety of surrogate or proxy measures are often used. These include the species richness of specific taxa, the number of distinct plant functional types (such as grasses, forbs, bushes, or trees), or the diversity of distinct gene sequences in a sample of microbial DNA taken from the soil. Species- or other taxon-based measures of biodiversity, however, rarely capture key attributes such as variability, function, quantity, and distribution—all of which provide insight into the roles of biodiversity.

Ecological indicators are scientific constructs that use quantitative data to measure aspects of biodiversity, ecosystem condition, services, or drivers of change, but no single ecological indicator captures all the dimensions of biodiversity. Ecological indicators form a critical component of monitoring, assessment, and decision-making and are designed to communicate information quickly and easily to policy-makers. In a similar manner, economic indicators such as GDP are highly influential and well understood by decision-makers. Some environmental indicators, such as global mean temperature and atmospheric CO_2 concentrations, are becoming widely accepted as measures of anthropogenic effects on global climate. Ecological indicators are founded on much the same principles and therefore carry with them similar pros and cons.

Biodiversity is typically a measure of variation at the genetic, species, and ecosystem level. Terrestrial biodiversity is usually greater near the equator, which is the result of the warm climate and high primary productivity. Biodiversity is not distributed evenly on Earth, and is richest in the tropics. These tropical forest ecosystems cover less than 10 percent of earth's surface, and contain about 90 percent of the world's species. Marine biodiversity is usually highest along coasts in the Western Pacific, where sea surface temperature is highest, and in the mid-latitudinal band in all oceans. There are latitudinal gradients in species diversity. Biodiversity generally tends to cluster in hotspots, and has been increasing through time, but will be likely to slow in the future.

Rapid environmental changes typically cause mass extinctions. More than 99.9 percent of all species that ever lived on Earth, amounting to over five billion species, are estimated to be extinct. Estimates on the number of Earth's current species range from 10 million to 14 million, of which about 1.2 million have been documented and over 86 percent have not yet been described. More recently, in May 2016, scientists reported that 1 trillion species are estimated to be on Earth currently with only one-thousandth of one percent described. The total amount of related DNA base pairs on Earth is estimated at 5.0 x 1037 and weighs 50 billion tonnes. In comparison, the total mass of the biosphere has been estimated to be as much as 4 TtC (trillion tons of carbon). In July 2016, scientists reported identifying a set of 355 genes from the Last Universal Common Ancestor (LUCA) of all organisms living on Earth.

The age of the Earth is about 4.54 billion years. The earliest undisputed evidence of life on Earth dates at least from 3.5 billion years ago, during the Eoarchean Era after a geological crust started to solidify following the earlier molten Hadean Eon. There are microbial mat fossils found in 3.48 billion-year-old sandstone discovered in Western Australia. Other early physical evidence of a biogenic substance is graphite in 3.7 billion-year-old meta-sedimentary rocks discovered in Western Greenland. More recently, in 2015, "remains of biotic life" were found in 4.1 billion-year-old rocks in Western Australia. According to one of the researchers, "If life arose relatively quickly on Earth then it could be common in the universe."

Since life began on Earth, five major mass extinctions and several minor events have led to large and sudden drops in biodiversity. The Phanerozoic eon (the last 540 million years) marked a rapid growth in biodiversity via the Cambrian explosion—a period during which the majority of multicellular phyla first appeared. The next 400 million years included repeated, massive biodiversity losses classified as mass extinction events. In the Carboniferous, rainforest collapse led to a great loss of plant and animal life. The Permian–Triassic extinction event, 251 million years ago, was the worst; vertebrate recovery took 30 million years. The most recent, the Cretaceous–Paleogene extinction event, occurred 65 million years ago and has often attracted more attention than others because it resulted in the extinction of the non-avian dinosaurs.

A sampling of fungi collected during summer 2008 in Northern Saskatchewan mixed woods, near LaRonge is anexample regarding the species diversity of fungi. In this photo, there are also leaf lichens and mosses.

The period since the emergence of humans has displayed an ongoing biodiversity reduction and an accompanying loss of genetic diversity. Named the Holocene extinction, the reduction is caused primarily by human impacts, particularly habitat destruction. Conversely, biodiversity positively impacts human health in a number of ways, although a few negative effects are studied.

TYPES OF BIODIVERSITY

Biodiversity is a key measure of the health of any ecosystem, and of our entire planet. Every organism in an ecosystem, or biome, relies on other organisms and the physical environment. For example, plant and animal species need each other for food, and depend on the environment for water and shelter. Biodiversity describes how much variety an ecosystem has, in terms of resources and species, and also genetically within species. A more diverse ecosystem will have more resources to help it recover from famine, drought, disease or even the extinction of a species. There are several levels of biodiversity, each indicating how diverse the genes, species and resources are in a region.

Species Diversity

Every ecosystem contains a unique collection of species, all interacting with each other. Some ecosystems may have many more species than another. In some ecosystems, one species has grown so large that it dominates the natural community. When comparing the biodiversity of ecosystems, an ecosystem that has a large number of species, but no species greatly outnumbering the rest, would be considered to have the most species diversity. A large number of species can help an ecosystem recover from ecological threats, even if some species go extinct.

Genetic Diversity

Genetic diversity describes how closely related the members of one species are in a given ecosystem. In simple terms, if all members have many similar genes, the species has low genetic diversity. Because of their small populations, endangered species may have low genetic diversity due to inbreeding. This can pose a threat to a population if it leads to inheritance of undesirable traits or makes the species more susceptible to disease. Having high genetic diversity helps species adapt to changing environments.

Ecosystem Diversity

A region may have several ecosystems, or it may have one. Wide expanses of oceans or deserts would be examples of regions with low ecological diversity. A mountain area that has lakes, forests and grasslands would have higher biodiversity, in this sense. A region with several ecosystems may be able to provide more resources to help native species survive, especially when one ecosystem is threatened by drought or disease.

Functional Diversity

The way species behave, obtain food and use the natural resources of an ecosystem is known as functional diversity. In general a species-rich ecosystem is presumed to have high functional diversity, because there are many species with many different behaviors. Understanding an eco-system's functional diversity can be useful to ecologists trying to conserve or restore damaged it, because knowing the behaviors and roles of species can point to gaps in a food cycle or ecological niches that are lacking species.

Species Diversity

Species diversity is the number of different species that are represented in a given community (a dataset). The effective number of species refers to the number of equally abundant species needed to obtain the same mean proportional species abundance as that observed in the dataset of interest (where all species may not be equally abundant). Species diversity consists of three components: species richness, taxonomic or phylogenetic diversity and species evenness. Species richness is a simple count of species, taxonomic or phylogenetic diversity is the genetic relationship between different groups of species,whereas species evenness quantifies how equal the abundances of the species are.

Calculation of Diversity

Species diversity in a dataset can be calculated by first taking the weighted average of species proportional abundances in the dataset, and then taking the inverse of this. The equation is:

$$^{q}D = \cfrac{1}{\sqrt[q-1]{\displaystyle\sum_{i=1}^{S} p_i p_i^{q-1}}}$$

The denominator equals mean proportional species abundance in the dataset as calculated with the weighted generalized mean with exponent q - 1. In the equation, S is the total number of species (species richness) in the dataset, and the proportional abundance of the ith species is p_i. The proportional abundances themselves are used as weights. The equation is often written in the equivalent form:

$$^{q}D = \left(\sum_{i=1}^{S} p_i^{q} \right)^{1/(1-q)}$$

The value of q defines which kind of mean is used. q = 0 corresponds to the weighted harmonic mean, which is 1/S because the p_i values cancel out. q = 1 is undefined, except that the limit as q approaches 1 is well defined:

$$\lim_{q \to 1} {}^{q}D = \exp\left(-\sum_{i=1}^{S} p_i \ln p_i \right)$$

q = 2 corresponds to the arithmetic mean. As q approaches infinity, the generalized mean approaches the maximum p_i value. In practice, q modifies species weighting, such that increasing q increases the weight given to the most abundant species, and fewer equally abundant species are hence needed to reach mean proportional abundance. Consequently, large values of q lead to smaller species diversity than small values of q for the same dataset. If all species are equally abundant in the dataset, changing the value of q has no effect, but species diversity at any value of q equals species richness.

Negative values of q are not used, because then the effective number of species (diversity) would exceed the actual number of species (richness). As q approaches negative infinity, the generalized mean approaches the minimum p_i value. In many real datasets, the least abundant species is represented by a single individual, and then the effective number of species would equal the number of individuals in the dataset.

The same equation can be used to calculate the diversity in relation to any classification, not only species. If the individuals are classified into genera or functional types, p_i represents the proportional abundance of the ith genus or functional type, and qD equals genus diversity or functional type diversity, respectively.

Diversity Indices

Often researchers have used the values given by one or more diversity indices to quantify species diversity. Such indices include species richness, the Shannon index, the Simpson index, and the complement of the Simpson index (also known as the Gini-Simpson index).

When interpreted in ecological terms, each one of these indices corresponds to a different thing, and their values are therefore not directly comparable. Species richness quantifies the actual rather than effective number of species. The Shannon index equals log(^qD), and in practice quantifies the uncertainty in the species identity of an individual that is taken at random from the dataset. The Simpson index equals $1/^qD$ and quantifies the probability that two individuals taken at random from the dataset (with replacement of the first individual before taking the second) represent the same species. The Gini-Simpson index equals $1 - 1/^qD$ and quantifies the probability that the two randomly taken individuals represent different species.

Sampling Considerations

Depending on the purposes of quantifying species diversity, the data set used for the calculations can be obtained in different ways. Although species diversity can be calculated for any data-set where individuals have been identified to species, meaningful ecological interpretations require that the dataset is appropriate for the questions at hand. In practice, the interest is usually in the species diversity of areas so large that not all individuals in them can be observed and identified to species, but a sample of the relevant individuals has to be obtained. Extrapolation from the sample to the underlying population of interest is not straightforward, because the species diversity of the available sample generally gives an underestimation of the species diversity in the entire population. Applying different sampling methods will lead to different sets of individuals being observed for the same area of interest, and the species diversity of each set may be different. When a new individual is added to a dataset, it may introduce a species that was not yet represented. How much this increases species diversity depends on the value of q: when q = 0, each new actual species causes species diversity to increase by one effective species, but when q is large, adding a rare species to a dataset has little effect on its species diversity.

In general, sets with many individuals can be expected to have higher species diversity than sets with fewer individuals. When species diversity values are compared among sets, sampling efforts need to be standardised in an appropriate way for the comparisons to yield ecologically meaningful results. Resampling methods can be used to bring samples of different sizes to a common footing. Species discovery curves and the number of species only represented by one or a few individuals can be used to help in estimating how representative the available sample is of the population from which it was drawn.

Genetic Diversity

Genetic diversity is the total number of genetic characteristics in the genetic makeup of a species. It is distinguished from genetic variability, which describes the tendency of genetic characteristics to vary.

Genetic diversity serves as a way for populations to adapt to changing environments. With more variation, it is more likely that some individuals in a population will possess variations of alleles that are suited for the environment. Those individuals are more likely to survive to produce off-spring bearing that allele. The population will continue for more generations because of the success of these individuals.

The academic field of population genetics includes several hypotheses and theories regarding genetic diversity. The neutral theory of evolution proposes that diversity is the result of the accumulation of neutral substitutions. Diversifying selection is the hypothesis that two subpopulations of a species live in different environments that select for different alleles at a particular locus. This may occur, for instance, if a species has a large range relative to the mobility of individuals within it. Frequency-dependent selection is the hypothesis that as alleles become more common, they become more vulnerable. This occurs in host–pathogen interactions, where a high frequency of a defensive allele among the host means that it is more likely that a pathogen will spread if it is able to overcome that allele.

Within Species Diversity

A study conducted by the National Science Foundation in 2007 found that genetic diversity (within species diversity) and biodiversity are dependent upon each other — i.e. that diversity within a species is necessary to maintain diversity among species, and vice versa. According to the lead researcher in the study, Dr. Richard Lankau, "If any one type is removed from the system, the cycle can break down, and the community becomes dominated by a single species." Genotypic and phenotypic diversity have been found in all species at the protein, DNA, and organismal levels; in nature, this diversity is nonrandom, heavily structured, and correlated with environmental variation and stress.

Varieties of maize in the office of the Russian plant geneticist Nikolai Vavilov.

The interdependence between genetic and species diversity is delicate. Changes in species diversity lead to changes in the environment, leading to adaptation of the remaining species. Changes in genetic diversity, such as in loss of species, leads to a loss of biological diversity. Loss of genetic diversity in domestic animal populations has also been studied and attributed to the extension of markets and economic globalization.

Evolutionary Importance of Genetic Diversity

Adaptation

Variation in the populations gene pool allows natural selection to act upon traits that allow the population to adapt to changing environments. Selection for or against a trait can occur with changing environment – resulting in an increase in genetic diversity (if a new mutation is selected for and maintained) or a decrease in genetic diversity (if a disadvantageous allele is selected against). Hence, genetic diversity plays an important role in the survival and adaptability of a species. The capability of the population to adapt to the changing environment will depend on the presence of the necessary genetic diversity The more genetic diversity a population has, the more likelihood the population will be able to adapt and survive. Conversely, the vulnerability of a population to changes, such as climate change or novel diseases will increase with reduction in genetic diversity. For example, the inability of koalas to adapt to fight Chlamydia and the koala retrovirus (KoRV) has been linked to the koala's low genetic diversity. This low genetic diversity also has geneticists concerned for the koalas ability to adapt to climate change and human-induced environmental changes in the future.

Small Populations

Large populations are more likely to maintain genetic material and thus generally have higher genetic diversity. Small populations are more likely to experience the loss of diversity over time by random chance, which is called genetic drift. When an allele (variant of a gene) drifts to fixation, the other allele at the same locus is lost, resulting in a loss in genetic diversity. In small population sizes, inbreeding, or mating between individuals with similar genetic makeup, is more likely to occur, thus perpetuating more common alleles to the point of fixation, thus decreasing genetic diversity. Concerns about genetic diversity are therefore especially important with large mammals due to their small population size and high levels of human-caused population effects.

A genetic bottleneck can occur when a population goes through a period of low number of individuals, resulting in a rapid decrease in genetic diversity. Even with an increase in population size, the genetic diversity often continues to be low if the entire species began with a small population, since beneficial mutations are rare, and the gene pool is limited by the small starting population. This is an important consideration in the area of conservation genetics, when working toward a rescued population or species that is genetically-healthy.

Mutation

Random mutations consistently generate genetic variation. A mutation will increase genetic diversity in the short term, as a new gene is introduced to the gene pool. However, the persistence of this gene is dependent of drift and selection. Most new mutations either have a neutral or negative effect on fitness, while some have a positive effect. A beneficial mutation is more likely to persist and thus have a long-term positive effect on genetic diversity. Mutation rates differ across the genome, and larger populations have greater mutation rates. In smaller populations a mutation is less likely to persist because it is more likely to be eliminated by drift.

Gene Flow

Gene flow, often by migration, is the movement of genetic material (for example by pollen in the wind, or the migration of a bird). Gene flow can introduce novel alleles to a population. These alleles can be integrated into the population, thus increasing genetic diversity.

For example, an insecticide-resistant mutation arose in Anopheles gambiae African mosquitoes. Migration of some A. gambiae mosquitoes to a population of Anopheles coluzziin mosquitoes resulted in a transfer of the beneficial resistance gene from one species to the other. The genetic diversity was increased in A. gambiae by mutation and in A. coluzziin by gene flow.

In Agriculture

In Crops

When humans initially started farming, they used selective breeding to pass on desirable traits of the crops while omitting the undesirable ones. Selective breeding leads to monocultures: entire farms of nearly genetically identical plants. Little to no genetic diversity makes crops extremely susceptible to widespread disease; bacteria morph and change constantly and when a disease-causing bacterium changes to attack a specific genetic variation, it can easily wipe out vast quantities of the species. If the genetic variation that the bacterium is best at attacking happens to be that which humans have selectively bred to use for harvest, the entire crop will be wiped out.

The nineteenth-century Potato Famine in Ireland was in part caused by lack of biodiversity. Since new potato plants do not come as a result of reproduction, but rather from pieces of the parent plant, no genetic diversity is developed, and the entire crop is essentially a clone of one potato, it is especially susceptible to an epidemic. In the 1840s, much of Ireland's population depended on potatoes for food. They planted namely the "lumper" variety of potato, which was susceptible to a rot-causing oomycete called Phytophthora infestans. The fungus destroyed the vast majority of the potato crop, and left one million people to starve to death.

Genetic diversity in agriculture does not only relate to disease, but also herbivores. Similarly, to the above example, monoculture agriculture selects for traits that are uniform throughout the plot. If this genotype is susceptible to certain herbivores, this could result in the loss of a large portion of the crop. One way farmers get around this is through inter-cropping. By planting rows of unrelated, or genetically distinct crops as barriers between herbivores and their preferred host plant, the farmer effectively reduces the ability of the herbivore to spread throughout the entire plot.

In livestock

The genetic diversity of livestock species permits animal husbandry in a range of environments and with a range of different objectives. It provides the raw material for selective breeding programmes and allows livestock populations to adapt as environmental conditions change.

Livestock biodiversity can be lost as a result of breed extinctions and other forms of genetic erosion. As of June 2014, among the 8,774 breeds recorded in the Domestic Animal Diversity Information System (DAD-IS), operated by the Food and Agriculture Organization of the United Nations (FAO), 17 percent were classified as being at risk of extinction and 7 percent already extinct. There is now a Global Plan of Action for Animal Genetic Resources that was developed under the auspices of the Commission on Genetic Resources for Food and Agriculture in 2007, that provides a framework and guidelines for the management of animal genetic resources.

Awareness of the importance of maintaining animal genetic resources has increased over time. FAO has published two reports on the state of the world's animal genetic resources for food and agriculture, which cover detailed analyses of our global livestock diversity and ability to manage and conserve them.

Viral Implications

High genetic diversity in viruses must be considered when designing vaccinations. High genetic diversity results in difficulty in designing targeted vaccines, and allows for viruses to quickly evolve to resist vaccination lethality. For example, malaria vaccinations are impacted by high levels of genetic diversity in the protein antigens. In addition, HIV-1 genetic diversity limits the use of currently available viral load and resistance tests.

Coping with Low Genetic Diversity

Natural

The natural world has several ways of preserving or increasing genetic diversity. Among oceanic plankton, viruses aid in the genetic shifting process. Ocean viruses, which infect the plankton, carry genes of other organisms in addition to their own. When a virus containing the genes of one cell infects another, the genetic makeup of the latter changes. This constant shift of genetic makeup helps to maintain a healthy population of plankton despite complex and unpredictable environmental changes.

Photomontage of planktonic organisms.

Cheetahs are a threatened species. Low genetic diversity and resulting poor sperm quality has made breeding and survivorship difficult for cheetahs. Moreover, only about 5% of cheetahs survive to adulthood However, it has been recently discovered that female cheetahs can mate with more than one male per litter of cubs. They undergo induced ovulation, which means that a new egg is produced every time a female mates. By mating with multiple males, the mother increases the genetic diversity within a single litter of cubs.

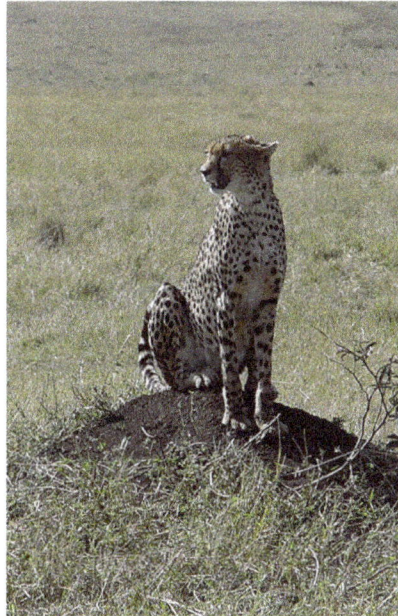
A Tanzanian cheetah.

Human Intervention

Attempts to increase the viability of a species by increasing genetic diversity is called genetic rescue. For example, eight panthers from Texas were introduced to the Florida panther population, which was declining and suffering from inbreeding depression. Genetic variation was thus increased and resulted in a significant increase in population growth of the Florida Panther. Creating or maintaining high genetic diversity is an important consideration in species rescue efforts, in order to ensure the longevity of a population.

Measures

Genetic diversity of a population can be assessed by some simple measures:

- Gene diversity is the proportion of polymorphic loci across the genome.

- Heterozygosity is the fraction of individuals in a population that are heterozygous for a particular locus.

- Alleles per locus is also used to demonstrate variability.

- Nucleotide diversity is the extent of nucleotide polymorphisms within a population, and is commonly measured through molecular markers such as micro- and minisatellite sequences, mitochondrial DNA, and single-nucleotide polymorphisms (SNPs).

Furthermore, stochastic simulation software is commonly used to predict the future of a population given measures such as allele frequency and population size.

Ecosystem Diversity

Ecosystem diversity deals with the variations in ecosystems within a geographical location and its overall impact on human existence and the environment.

Ecosystem diversity is a type of biodiversity. It is the variation in the ecosystems found in a region or the variation in ecosystems over the whole planet. Biodiversity is important because it clears out our water, changes out climate, and provides us with food. Ecological diversity includes the variation in both terrestrial and aquatic ecosystems. Ecological diversity can also take into account the variation in the complexity of a biological community, including the number of different niches, the number of trophic levels and other ecological processes. An example of ecological diversity on a global scale would be the variation in ecosystems, such as deserts, forests, grasslands, wetlands and oceans. Ecological diversity is the largest scale of biodiversity, and within each ecosystem, there is a great deal of both species and genetic diversity.

The Earth has many diverse ecosystems and ecologicalsystem diversity.

Diversity in the ecosystem is significant to human existence for a variety of reasons. Ecosystem diversity boosts the availability of oxygen via the process of photosynthesis amongst plant organisms domiciled in the habitat. Diversity in an aquatic environment helps in the purification of water by plant varieties for use by humans. Diversity increases plant varieties which serves as a good source for medicines and herbs for human use. A lack of diversity in the ecosystem produces an opposite result.

Some examples of ecosystems that are rich in diversity are:

- Deserts
- Forests
- Large marine ecosystems
- Marine ecosystems
- Old growth forests
- Rainforests
- Tundra
- Coral Reefs
- Marine

Ecosystem Diversity as a Result of Evolutionary Pressure

Ecological diversity around the world can be directly linked to the evolutionary and selective pressures that constrain the diversity outcome of the ecosystems within different niches. Tundras, Rainforests, coral reefs and deciduous forests all are formed as a result of evolutionary pressures. Even seemingly small evolutionary interactions can have large impacts on the diversity of the ecosystems throughout the world. One of the best studied cases of this is of the honey bee's interaction with angiosperms on every continent in the world except Antarctica.

In 2010 Robert Brodschneider, and Karl Crailsheim conducted a study about the health and nutrition in honey bee colonies, the study conducted focused on: overall colony health, adult nutrition, and larva nutrition as a function of the effect of pesticides, monocultures and genetically modified crops to see if the anthropogenically created problems can have an effect pollination levels. The results indicate that human activity does have a role in the destruction of the fitness of the bee colony. The extinction or near extinction of these pollinators would result in many plants that feed humans on a wide scale, needing alternative pollination methods. Crop pollinating insects are worth annually 14.6 billion to the US economy and cost to hand pollinate over insect pollination will cost an estimated 5,715-$7,135 per hectare additionally. Not only will there be a cost increase but also an decrease in colony fitness, leading to a decrease in genetic diversity, which studies have shown has a direct link to the long term survival of the honey bee colonies.

According to a study, there are over 50 plants that are dependent on bee pollination, many of these being key staples to feeding the world. Another study conducted states that as a direct result of a lack of plant diversity, will lead to a decline in the bee population fitness, and a low bee colony fitness has impacts on the fitness of plant ecosystem diversity. By allowing for bee pollination and working to reduce anthropogenically harmful footprints, bee pollination can increase flora growth genetic diversity and create a unique ecosystem that is highly diverse and can provide a habitat and niche for many other organisms to thrive. Due to the evolutionary pressures of bees being located on six out of seven continents, there can be no denying the impact of pollinators on the ecosystem diversity. The pollen collected by the bees is harvested and used as an energy source for winter time, this act of collecting pollen from local plants also has a more important effect of facilitating the movement of genes between organisms.

The new evolutionary pressures that are largely anthropogenically catalyzed can potentially cause wide spread collapse of ecosystems. In the north Atlantic sea, a study was conducted that followed the effects of the human interaction on surrounding ocean habitats. They found that in there was no habitat or trophic level that in some way was effected negatively by human interaction, and that much of the diversity of life was being stunted as a result.

Functional Diversity (Disability)

Functional diversity is a politically and socially correct term for special needs, disability, impairment and handicap, which began to be used in Spain in scientific writing, at the initiative of those directly affected, in 2005.

Usage

This term is intended to replace other ones with pejorative semantics. It proposes a shift towards non-negative, non-disparaging and non-patronizing terms. The formal justification of the term can be found in the book El Modelo de la Diversidad by Agustina Palacios and Javier Romañach, Examples of usage:

- "People with functional diversity" instead of "people with special needs".

- "Physical functional diversity".

- "Mobility functional diversity", "person who uses a wheelchair", "wheelchair user".

- "Motor functional diversity".

- "Dexterity functional diversity".

- "Visual functional diversity", "people who use screen readers as their primary means of accessing a computer".

- "People with a visual processing functional diversity".

- "Auditory functional diversity".

- "Mental functional diversity".

- "Intellectual functional diversity".

- "Cognitive functional diversity".

- "Organic functional diversity".

- "Circumstantial and/or temporary functional diversity".

- "Person with a functional diversity".

- "Persons without functional diversity" rather than "normal" or "healthy".

- "People without functional diversity", "typically developing children".

MEASURING BIODIVERSITY

Just as there are many different ways to define biodiversity, there are many different measures of biodiversity. Most measures quantify the number of traits, individuals, or species in a given area while taking into account their degree of dissimilarity. Some measure biodiversity on a genetic level while others measure diversity within a single habitat or between ecosystems.

Measuring biodiversity on the genetic level requires that researchers map the genes and chromosomes of an individual organism and then compare them to the genetic make-up of the larger population. It is genetic diversity which causes tulips to be different colors and different heights. Typically, researchers measure genetic diversity by counting how often certain genetic patterns occur. Another method of measuring genetic diversity works in the reverse: researchers evaluate the differences in physical appearance between individuals then attribute these traits to the most likely genetic roots. Mapping diversity at the genetic level is currently the most accurate measure of biodiversity, although it can be costly and time consuming and, thus, impractical for evaluating large ecosystems. It is most often used to examine managed populations or agricultural crops which can allow for selective breeding of the most desirable traits.

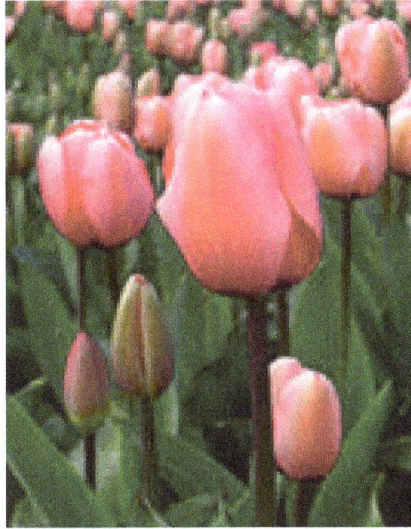

Measuring the diversity of a species generally incorporates estimates of "richness." Also referred to as alpha-diversity, species richness is a common way of measuring biodiversity and involves counting the number of individuals – or even families – in a given area. Researchers have created several indices which measure species biodiversity, the most popular are the Simpson Index and the Shannon Index. These indices focus on the relative species richness and abundance and/or the pattern of species distribution. The Simpson Index takes into account the number of species present and their relative abundance in proportion to the total population. The Shannon Index, originally developed for use in information science, accounts for the order or abundance of a species within a sample plot. The Shannon Index is often used for identifying areas of high natural or human disturbance.

There are also many challenges when measuring species diversity. The greatest of which is a lack of available data. Conducting a full count of the number of species in an ecosystem is nearly impossible, so researchers must use sample plots at a variety of sites but must avoid repetitive counting. Oftentimes, information is not compiled in one specific place, a problem that can lead to an overlap in the naming of species. Another limitation is an inconsistency in treating the definition of species: what one scientist may classify as a new species another may not.

At the ecosystem-level, measures of biodiversity are often used to compare two ecosystems or to determine changes over time in a given region. Describing changes in biodiversity within or between ecosystems is called beta-diversity. Measures of beta-diversity indicate the difference in species richness between two different habitats or within a single community at different points in time. The resulting number indicates to researchers whether there is any overlap in the species found in each group. Gamma-diversity, on the other hand, estimates the total biodiversity within an entire region. To arrive at a total estimate, researchers may set up sample plots around the region and count all species within the plots. The sizes of the plots can vary depending on the physical characteristics of the locale. For example, plots in northern forests may be as large as a hectacre whereas in dense rainforest a plot might only be a few meters. Another indicator of biodiversity which researchers often track and measure are keystone species, which are integral to ecosystem processes.

Measuring biodiversity on an ecosystem level is thought to be a better way of looking at the health of the entire system, rather than the health of a particular species. However, it faces many of the same challenges measuring species and genetic diversity do – primarily in cost and the lack of standardization. Researchers have only begun taking measurements; this further limits their ability to identify trends since ecosystems tend to change slowly over time. This absence of long-term scientific data remains a particular challenge.

Counting animals and plants, mapping genes, and systematically comparing ecosystems may seem like a lot of trouble for a number that is – ultimately – an estimate. However, the numbers matter. In the field of conservation, biodiversity is often a consideration within an area; being able to quantify what is being conserved is imperative to good planning and management. Labeling a species or ecosystem "diverse" becomes relative; an estimate of biodiversity will have recognizable limitations, like those of imperfect sampling, but will give a comparison or point of reference. The creation of indices gives scientists a standardized tool with which to compare both ecosystem and species health. Therefore, although exact diversity numbers are difficult to yield, knowing how biological resources are distributed within a community can be extremely beneficial in determining both short- and long-term trends.

DISTRIBUTION OF BIODIVERSITY

Biodiversity is not evenly distributed, rather it varies greatly across the globe as well as within regions. Among other factors, the diversity of all living things (biota) depends on temperature, precipitation, altitude, soils, geography and the presence of other species. The study of the spatial distribution of organisms, species and ecosystems, is the science of biogeography.

Diversity consistently measures higher in the tropics and in other localized regions such as the Cape Floristic Region and lower in polar regions generally. Rain forests that have had wet climates for a long time, such as Yasuní National Park in Ecuador, have particularly high biodiversity.

Terrestrial biodiversity is thought to be up to 25 times greater than ocean biodiversity. A new method used in 2011, put the total number of species on Earth at 8.7 million, of which 2.1 million were estimated to live in the ocean. However, this estimate seems to under-represent the diversity of microorganisms.

A conifer forest in the Swiss Alps (National Park).

Latitudinal Gradients

Generally, there is an increase in biodiversity from the poles to the tropics. Thus localities at lower latitudes have more species than localities at higher latitudes. This is often referred to as the latitudinal gradient in species diversity. Several ecological factors may contribute to the gradient, but the ultimate factor behind many of them is the greater mean temperature at the equator compared to that of the poles.

Even though terrestrial biodiversity declines from the equator to the poles, some studies claim that this characteristic is unverified in aquatic ecosystems, especially in marine ecosystems. The latitudinal distribution of parasites does not appear to follow this rule.

In 2016, an alternative hypothesis ("the fractal biodiversity") was proposed to explain the biodiversity latitudinal gradient. In this study, the species pool size and the fractal nature of ecosystems were combined to clarify some general patterns of this gradient. This hypothesis considers temperature, moisture, and net primary production (NPP) as the main variables of an ecosystem niche and as the axis of the ecological hypervolume. In this way, it is possible to build fractal hypervolumes, whose fractal dimension rises up to three moving towards the equator.

Hotspots

A biodiversity hotspot is a region with a high level of endemic species that have experienced great habitat loss. The term hotspot was introduced in 1988 by Norman Myers. While hotspots are spread all over the world, the majority are forest areas and most are located in the tropics.

Brazil's Atlantic Forest is considered one such hotspot, containing roughly 20,000 plant species, 1,350 vertebrates and millions of insects, about half of which occur nowhere else. The island of Madagascar and India are also particularly notable. Colombia is characterized by high biodiversity, with the highest rate of species by area unit worldwide and it has the largest number of endemics (species that are not found naturally anywhere else) of any country. About 10% of the species of the Earth can be found in Colombia, including over 1,900 species of bird, more than in Europe and North America combined, Colombia has 10% of the world's mammals species, 14% of the amphibian species and 18% of the bird species of the world. Madagascar dry deciduous forests and lowland rainforests possess a high ratio of endemism. Since the island separated from mainland Africa 66 million years ago, many species and ecosystems have evolved independently. Indonesia's 17,000 islands cover 735,355 square miles (1,904,560 km²) and contain 10% of the world's flowering plants, 12% of mammals and 17% of reptiles, amphibians and birds—along with nearly 240 million people. Many regions of high biodiversity and/or endemism arise from specialized habitats which require unusual adaptations, for example, alpine environments in high mountains, or Northern European peat bogs.

Accurately measuring differences in biodiversity can be difficult. Selection bias amongst researchers may contribute to biased empirical research for modern estimates of biodiversity. In 1768, Rev. Gilbert White succinctly observed of his Selborne, Hampshire "all nature is so full, that that district produces the most variety which is the most examined."

Because of the curvature of the Earth and the fact that it is tilted slightly on its axis relative to the sun, different regions of the planet receive different amounts of sunlight energy throughout the year. This impacts the length of warm, cold, wet, and dry seasons in these different regions, as well as the temperature, humidity, and other environmental factors that define the region. A biodiversity hotspot is a region containing an exceptional concentration of endemic species, but is threatened by human-induced loss of habitat. These hot spots support nearly 60% of the world's plant, bird, mammal, reptile, and amphibian species. Many global organizations are working to conserve these biodiversity hotspots, such as the World Wildlife Foundation's Global 200 and the Critical Ecosystem Partnership Fund. A relatively small number of countries (17) have less than 10% of the global surface, but support more than 70% of the biological diversity on earth. Countries rich in biological diversity and associated traditional knowledge belong to a group known as the Like Minded Megadiverse Countries (LMMC).

Another consequence if the Earth's curvature and rotation is that the hydrologic cycle distributes water differently among these different regions. The result is striking differences in the global distribution of rain and snow. As a result, different regions on the planet have specific sets of environmental conditions, which results in differences in predominant vegetation. Species residing in different regions are characterized by specific adaptations that allow success under the particular set of environmental conditions of the region. Regions can be broadly divided into terrestrial biomes and aquatic ecosystems.

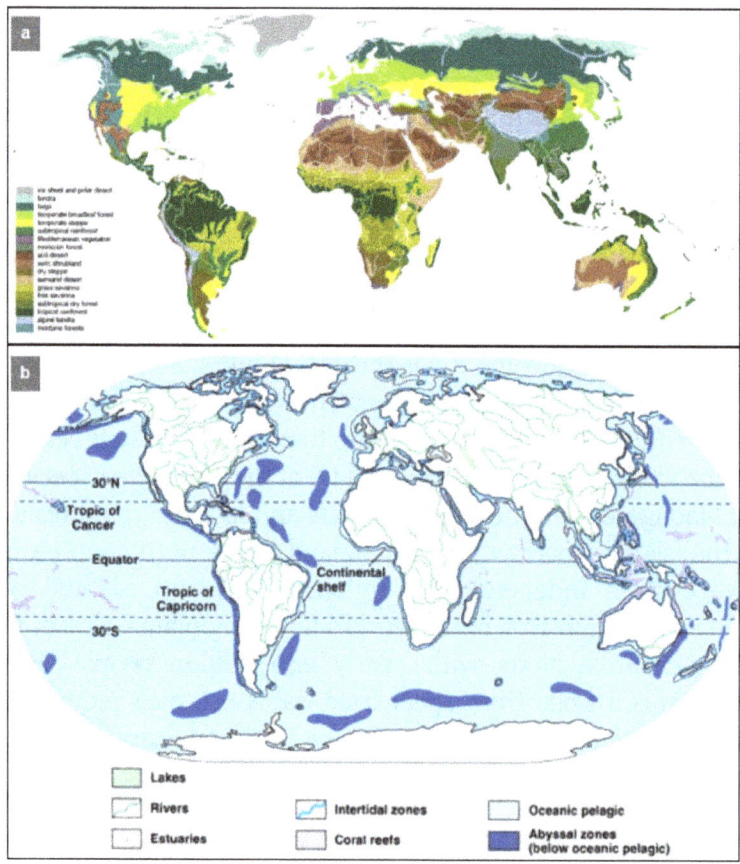

Distribution of: a. terrestrial biomes; b. aquatic ecosystems.

Terrestrial Biomes

The terrestrial biomes can be divided into four broad categories: forest, desert, savanna/grassland, and tundra.

Forests

Forest biomes are dominated by trees. Approximately one-third of Earth's land area is covered by forests that contain 70% of the carbon present in living things. Forest biomes are extremely important in buffering climate change since they remove carbon dioxide from the atmosphere during photosynthesis.

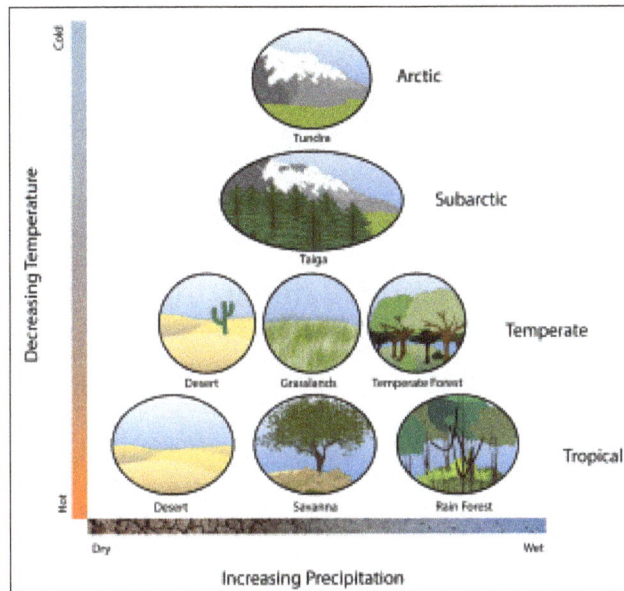

Characteristics of terrestrial biomes based on temperature and precipitation.

Forest biomes can be divided into three distinct types based primarily on the types of organisms that populate them and seasonal changes in temperature and/or precipitation. These three types are tropical, temperate, and boreal.

Seven terrestrial biomes: a. savanna; b. temperate grassland; c. and d. desert;
e. tropical forest; f. temperate deciduous forest; g. boreal forest; h. tundra.

Tropical forests support the highest biodiversity of all biomes. They occur near the equator where 1) day lengths are long and vary little from 12 hours, 2) rainfall is higher than any other biome, and 3) temperatures are high, averaging around 20-25° C, with little seasonal variation.

Deforestation is a significant problem in this biome and is occurring rapidly for a number of reasons: logging of particular tree species, like teak and mahogany for fine furniture; clearing land for farming or cattle production; oil drilling and mining; and establishment of plantations such as those for palm oil or sugar cane.

Temperate forests are typically dominated by deciduous tree species, which lose their leaves every autumn. These forests also support pines, hemlocks, and other conifers. The location of temperate forests is in the mid-latitudes (between 30°N and 45°N and latitudes 30°S and 45°S). In these latitudes, forests experience four well-defined seasons. Precipitation (75-150 cm) is distributed evenly throughout the year.

Deforestation is a problem in this biome as well, with much of the primary, old-growth temperate forests cleared by humans for fuel wood, building materials, and as a source of wood pulp to make paper. As such, many of the land covered in deciduous forest in the United States and elsewhere today contain regrown, secondary forests.

Boreal forests, also called Taiga, are dominated by coniferous, cone-bearing trees. The needles on these trees remain green throughout the year. This biome covers extensive swaths of land between 50 - 60° north latitudes. In these latitudes, seasons are divided into short, and moderately warm summers and long, cold winters. Winter temperatures are very low, with snow contributing the most to annual precipitation, which is 40-100 cm annually.

Boreal forests in Russia, Asia, and North America are currently threatened by climate change. For example, as a result of warming temperatures, infestation of native pest species like the mountain pine beetle are on the rise because soils in which these beetles occur during winter no longer freeze. Thus, an important natural control of these pests has been lost and their populations have increased greatly, causing deforestation to the trees and a cascading negative effect on the Taiga food web.

Stands of lodgepole pine killed by damage from the mountain pine beetle.

Deserts

Deserts cover about one fifth of the Earth's land surface and occur where rainfall is less than 50 cm a year. These are the driest landscapes on Earth and support the least amount of life. Biodiversity is lowest in these biomes. Most deserts occur along latitudes of 30 °N and 30 °S and therefore have generally hot climates. These regions receive little precipitation due to atmospheric circulation patterns.

Deserts can also occur at other latitudes and are produced in different ways. Rain shadow deserts are found on the leeward side of large mountain ranges. In these cases, warm moist air coming off oceans with the prevailing winds hits a mountain range and is deflected upwards. As the air rises, it cools and drops its moisture on the windward side of the range, creating cool, dry air after passing over the mountains; dry enough to form a desert region. Good examples of deserts formed by rain shadows are the Great Basin Desert on the leeward side of the Sierra Nevada mountain range in the western U.S., and the Patagonian Desert on the leeward side of the Andes in South America.

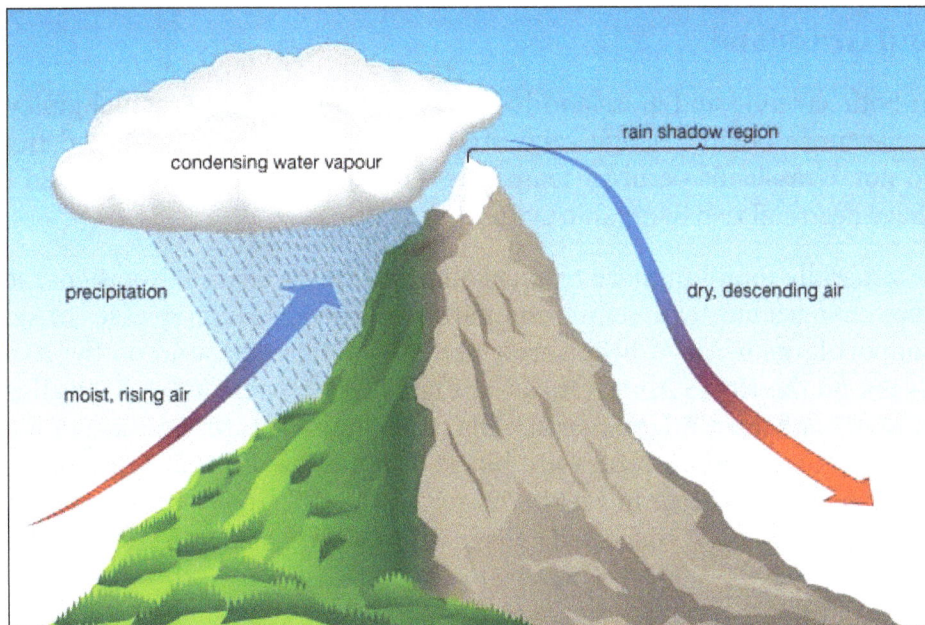

Rain shadow effect can form deserts on the
down-wind side of coastal mountain ranges.

Coastal deserts are found on the southwest coasts of South America (Atacama Desert) and Africa (Namib Desert) where deep, cold upwelling currents of water rise to the ocean surface and cool the air directly above, causing the air to release its moisture through precipitation over the ocean before it reaches the land. Continental interior deserts occur in areas in the interiors of large continents that are extremely far from a source of moisture. They are the driest and most lifeless of all biomes; the Gobi and Sahara Deserts are examples.

Polar deserts occur at the north and south poles where dry cold air prevails. Northern Greenland and non-ice covered areas of Antarctica are examples of polar deserts. Most deserts have a considerable amount of specialized vegetation as well as specialized vertebrate and invertebrate animals.

Because water is a precious resource, the leaves of many desert plants have evolved into spines to reduce water loss from larger leaves and protect their moisture-filled stems from being eaten by animals. In the case of the teddy-bear cholla cactus the spines serve additional roles of reflecting intense sunlight and trapping moisture. The Kit Fox, native to the Sonoran Desert in the U.S. Southwest, uses its large ears for evaporative cooling.

Desert Biota: a. Teddy-bear cholla cactus; b. Kit Fox; c. Diamondback Rattlesnake.

Savanna and Grassland

Vegetation in both savanna and grassland biomes is dominated by perennial grasses and non-woody forbs. Savannas obtain enough rainwater annually to support scattered trees, whereas grasslands do not. Grasslands occur in temperate climates with hot summers and cold, snowy winters and have deep soil rich in organic matter.

Savannas are generally found in more tropical climates where seasonality is characterized not by temperature changes but by precipitation patterns. The abundant grasses of savannas and grasslands support large herds of herbivores, like the wildebeest found on the African savanna and the bison on the North American Great Plains. Wildebeest, zebras, gazelles, and other large African mammals must migrate seasonally, tracking moisture. Because of human settlement, however, many migration routes are blocked by fences or other types of development. The American bison used to occur in vast herds across the U.S. grasslands but was nearly hunted to extinction in the 1800s. A concerted conservation effort has put this species back on track to recovery.

Wildebeest on the African Savanna; b. Bison on the American Great Plains.

Because of the deep roots and quick growth of these grasses, the soils have become rich with organic carbon making them valuable for agriculture. Much of the planet's natural grassland biome has been converted to farmland, which has caused a loss of these rich, valuable soils and a decrease in biodiversity.

Tundra

Tundra is among the coldest biomes, with average winter temperatures of -34 °C and summer temperatures between 3-12 °C. The warmer growing season lasts only 50 – 60 days, but this is adequate to supply sustenance to its multitudes of migrating birds and caribou. Tundra soil is high in organic matter and lies atop permanently frozen soil called permafrost. With the increased temperatures induced by climate change, however, shallower layers of permafrost are beginning to thaw, allowing the organic content to decay, releasing methane (CH_4) into the atmosphere that had been sequestered in organic form for millennia. This enormous release of methane from thawing tundra contributes considerably to greenhouse gas emissions.

Migratory species that rely on the tundra biome during the short growing season: a. tundra swan, b. caribou.

Aquatic Ecosystems

Water is the common link among the aquatic ecosystems and it makes up the largest portion of the biosphere. This is where life began billions of years ago. Without water, organisms would be unable to sustain themselves. Aquatic ecosystems support highly diverse groups of organisms and are classified into two broad categories: freshwater and saltwater or marine.

Freshwater Ecosystems

Freshwater ecosystems are characterized by having a very low salt (NaCl) content (less than 0.5 parts salt per 1,000 parts H2o, ppt) and include streams/rivers, groundwater, lakes, ponds, reservoirs, and wetlands (such as fens, marshes, swamps, and bogs). Each presents unique conditions to which different kinds of organisms are adapted. Life in flowing water (called lotic systems), for example, requires different adaptations than life in ponds, lakes, reservoirs, and wetlands (still water or lentic systems).

Because climatic conditions vary across different latitudes, the species diversity in freshwater aquatic ecosystems differs geographically. Like terrestrial biomes, aquatic ecosystems in the tropics support many more species than those in latitudes further from the equator. This is particularly true for fish and amphibians. The Amazon River, for example, which runs on or near the equator supports 2,000 to greater than 5,000 species of fish. This is a very high fish diversity. By contrast, the Mississippi River basin in the United States, which runs from approximately 45° N to 30° N latitude (from the headwaters to the mouth of the river), harbors only about 375 species of fish. Collectively, approximately 15,000 of the earth's species of fish, nearly 45% of all fish species, rely on either fresh or brackish water habitats. The other 55% are marine species.

Biotic Diversity in Freshwaters Vascular Plants [a-d] a. water lily, b. many-head rush, c. foxtail; Mirophyllum- d. bladderwort ; Ultricularia- Microalgae [e-h]: e. Green Alga Stigeoclonium, f. Green Algae - Micrasterias, Staurastrum & Xanthidium, g. Diatom – Cymbella, h. Cyanobacterium – Lyngbya; Protozoa & Metazoa [i – l]: i. Amoeba, j. Rotifer - Habrotrocha, k. Cladoceran – Bosmina, l. Copepod – Cyclops; Macroinvertebrates [m – p]: m. mayfly nymph, n. Diving bell spider, o. damselfly larvae, p. amphipod from groundwater; Fish [q – t]: q. Brook Trout, r. mudskipper, s. Neon tetra, t. American Paddlefish; Other Vertebrates [u – x]: u. Broad-Shelled River Turtle, v. Green Pygmy Goose, w. Orange-thighed tree frog, x. Asian small-clawed otter.

Microalgae both benthic (living on the bottom) and planktonic (living in the water column), are the major primary producers in most aquatic ecosystems and thus serve as the base of the food chain. As with all organismal groups, the number and identity of algal species in a community can influence ecosystem processes at a much larger scale. When the base of a food chain is diverse, higher trophic levels tend to be diverse as well.

Marine Ecosystems

Marine ecosystems contain salt that eroded from land and eventually washed into the oceans. The average ocean salinity is 35 ppt worldwide. Marine ecosystems cover about three-fourths of the Earth's surface and include oceans, seas, coral reefs, and estuaries. Estuaries are wetlands at the oceans' shore that contain a mix of freshwater from rivers and saltwater from the ocean to produce brackish water, characterized by possessing a salinity between 0.5 ppt and 17 ppt. Marine phytoplankton are critical to all life on Earth because they supply much of the world's atmospheric oxygen and take in a huge amount of atmospheric carbon dioxide for photosynthesis, acting as a "sink" for the greenhouse gas CO_2.

There are six distinct marine eco-regions. All of these, like terrestrial biomes, are characterized by specific flora and fauna.

Estuaries

Marine ecosystems: a. estuaries, b. intertidal, c. sub-tidal, d. coral reef, e. pelagic, f. abyssal.

Estuaries are formed at the mouths of freshwater streams or rivers flowing into the ocean. Depending on the elevation gradient of the land and the ratio of water flow from river to ocean versus intrusion from ocean to river, estuaries can range in salinity from 0.5 ppt to 17 ppt. This mixing of waters with such different salt and nutrient concentrations creates a very rich and unique ecosystem at the edge of two very different aquatic systems. The blending of two distinct systems at their border is called ecotone and is often a zone of high biodiversity because it harbors species from both systems. Estuaries have higher diversity and productivity than either the river or stream alone. Microflora like algae, and macroflora, such as seaweeds, marsh grasses, and mangrove trees (only in the tropics), can be found here. Estuaries support a diverse fauna, including a variety of worms, oysters, crabs, and waterfowl, and are often important nursery grounds for fish and important feeding stops for migratory birds.

Intertidal and Sub-Tidal Zones

Marine ecosystems along the coasts of land masses, but not influenced by infusion of freshwater like estuaries, include the intertidal and sub-tidal zones. Intertidal ecosystems are alternately exposed to the air and submerged as ocean tides wax and wane. Most species that live in this ecosystem are tolerant to and often thrive on periodic exposure to air, like mussels, crabs, starfish, sea anemones, and seaweeds. Tide pools, small shoreline depressions that retain permanent water, can even support a diversity of fish.

Sub-tidal zones occur further offshore and are permanently submerged but still strongly influenced by tidal surges. Dense kelp forests or sea-grass beds can grow in these areas, serving as habitat for multitudes of fish, shrimp, and other marine organisms.

Coral Reefs, Sea Grass Beds and Mangroves

Coral reefs are some of the most highly diverse ecosystems on Earth. They are widely distributed in warm shallow ocean waters. They can be found as barriers along continents, fringing islands, and atolls. Naturally, the dominant organisms in coral reefs are corals. Corals are interesting since they consist of a symbiosis between algae (zooxanthellae) and animal polyps housed with a calcareous, shell-like structure. Since coral reef waters tend to be nutritionally poor, Zooxanthellae the animal coral polyps obtain their nutrients through the algae via photosynthesis (where glucose is produced) and also by extending tentacles to obtain and ingest plankton from the water. The calcareous structure of the coral provides important habitat for a wide diversity of small and very colorful fish species, most of which live only on coral reefs.

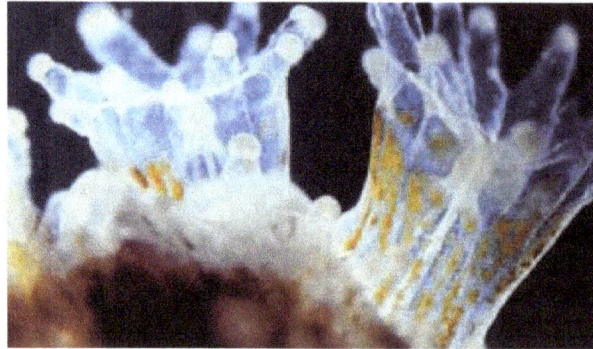

Algal zooxanthellae (brown dots) within coral polyps.

Coral reefs and adjacent sea-grass beds and mangrove forests are of high economic and ecological value to tropical countries, but at the same time are very sensitive to environmental changes, both natural and anthropogenic.

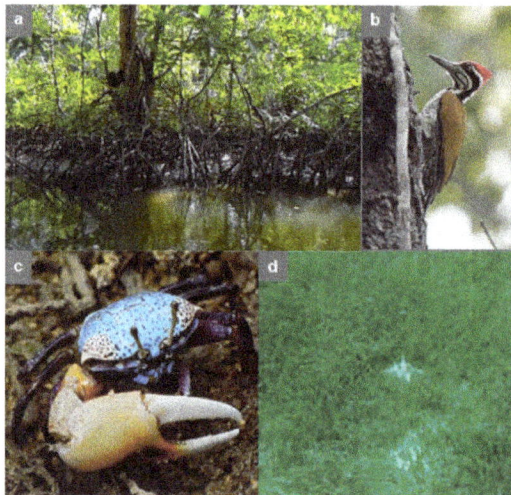

a. Mangrove forest, b. Greater flameback, c. Blue fiddler crab, d. seagrass bed.

Coral reefs, for example, act as barriers, protecting sea grasses and mangroves from oceanic swell and storms but are vulnerable to harm by tourists diving and collecting coral and by the aquarium industry that collects millions of colorful fish to sell on the market.

The region supporting the world's highest marine biodiversity is what has come to be known as the "Coral Triangle". This region encompasses parts of Southeast Asia and the western Pacific that surround Indonesia, Malaysia, Papua New Guinea, the Philippines, Timor Leste, and the Solomon Islands. Because this region is an extensive island archipelago, it has vast areas of shallow off-shore marine shelves that provide perfect conditions for coral reef formation. The Coral Triangle supports the greatest concentration of coral and reef fish species on earth. It has 76% of the world's 798 coral species, 37% of the world's 6000 coral reef fish species, and 56% of the 4050 coral reef fishes in the Indo-Pacific region.

Map of the Coral Triangle.

Major groups of marine species on the International Union for Conservation of Nature (IUCN) Red List of threatened species include the following:

- All the world's known species of reef-building corals (845 species).

- Sharks, rays and chimaera (1,046 species).

- Groupers (161 species).

- Seabirds (349 species).

- Marine mammals, which include whales, dolphins, porpoises, seals, sea lions, walruses, sea otter, marine otter, manatees, dugong and the polar bear (134 species).

- Marine turtles (7 species).

- Seagrasses and mangrove.

Pelagic Zone

Commercial catch of jack mackerel.

The Pelagic Zone comprises all far off-shore 'open water' habitat extending from the ocean surface to the depth limits of light penetration. This zone supports the massive schools of planktivorous forage fish like anchovies, smelt, and sardines, which serve as the primary diet of salmon, swordfish, tuna, and many other larger fish. Stocks of many of the large predatory fish are declining due to over fishing of both the forage fish and larger predatory fish themselves. Forage fish are subject to overfishing in areas where they are used to produce feed for farmed fish or commercial production of pet food.

Abyssal Zone

Hydrothermal vents occur at mid-ocean
ridges where new crust is formed.

The abyssal zone is the deepest region of oceans that lies below the pelagic zone, its upper limit at the depth where sunlight can no longer penetrate. Because these deep waters are in constant darkness, no photosynthetic organisms live there, yet a diversity of unique life still thrives comprising an unusually complex food web with bacteria, rather than microalgae, serving as the food web base.

In the complete darkness of the abyssal zone, predators, like the angler fish have developed evolutionary adaptations to allow them to capture prey. The angler fish uses a fluorescent lure extending off its head to attract prey.

The 'base of the food chain' or food that supports abyssal zone life comes from the bacteria that feeds on feces and the bodies of dead organisms raining down from the pelagic zone. In addition to these decomposer bacteria, other sea floor habitats promote life. At locations where molten magna emerges through the sea floor, creating new crust and pushing crustal plates apart, warm and nutrient rich water emerges from hydrothermal vents.

Angler Fish

These provide unique habitats for chemosynthetic bacteria to grow. Both decomposers and chemosynthetic bacteria provide a rich food source for a multitude of invertebrate fish species that are unique to the ocean's abyss.

BIOLOGICAL INTERACTION

The black walnut secretes a chemical from its roots that harms
neighboring plants, an example of competitive antagonism.

In ecology, a biological interaction is the effect that a pair of organisms living together in a community have on each other. They can be either of the same species (intraspecific interactions), or of different species (interspecific interactions). These effects may be short-term, like pollination and predation, or long-term; both often strongly influence the evolution of the species involved. A long-term interaction is called a symbiosis. Symbioses range from mutualism, beneficial to both partners, to competition, harmful to both partners. Interactions can be indirect, through intermediaries such as shared resources or common enemies.

Short-Term Interactions

Short-term interactions, including predation and pollination, are extremely important in ecology and evolution. These are short-lived in terms of the duration of a single interaction: a predator kills and eats a prey; a pollinator transfers pollen from one flower to another; but they are extremely durable in terms of their influence on the evolution of both partners. As a result, the partners coevolve.

Predation

Predation is a short-term interaction, in which the
predator, here an osprey, kills and eats its prey.

In predation, one organism, the predator, kills and eats another organism, its prey. Predators are adapted and often highly specialized for hunting, with acute senses such as vision, hearing, or smell.

Many predatory animals, both vertebrate and invertebrate, have sharp claws or jaws to grip, kill, and cut up their prey. Other adaptations include stealth and aggressive mimicry that improve hunting efficiency. Predation has a powerful selective effect on prey, causing them to develop antipredator adaptations such as warning coloration, alarm calls and other signals, camouflage and defensive spines and chemicals. Predation has been a major driver of evolution since at least the Cambrian period.

Pollination

In pollination, pollinators including insects (entomophily), some birds (ornithophily), and some bats, transfer pollen from a male flower part to a female flower part, enabling fertilisation, in return for a reward of pollen or nectar. The partners have coevolved through geological time; in the case of insects and flowering plants, the coevolution has continued for over 100 million years. Insect-pollinated flowers are adapted with shaped structures, bright colours, patterns, scent, nectar, and sticky pollen to attract insects, guide them to pick up and deposit pollen, and reward them for the service. Pollinator insects like bees are adapted to detect flowers by colour, pattern, and scent, to collect and transport pollen (such as with bristles shaped to form pollen baskets on their hind legs), and to collect and process nectar (in the case of honey bees, making and storing honey). The adaptations on each side of the interaction match the adaptations on the other side, and have been shaped by natural selection on their effectiveness of pollination.

Pollination has driven the coevolution of flowering plants and their animalpollinators for over 100 million years.

Symbiosis: Long-Term Interactions

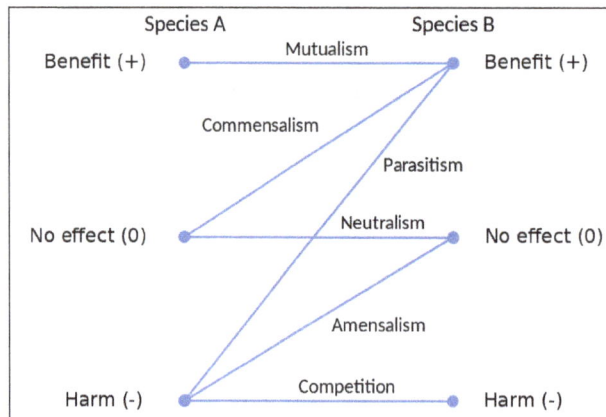

The six possible types of symbiotic relationship, from mutual benefit to mutual harm.

The six possible types of symbiosis are mutualism, commensalism, parasitism, neutralism, amensalism, and competition. These are distinguished by the degree of benefit or harm they cause to each partner.

Mutualism

Mutualism is an interaction between two or more species, where species derive a mutual benefit, for example an increased carrying capacity. Similar interactions within a species are known as co-operation. Mutualism may be classified in terms of the closeness of association, the closest being symbiosis, which is often confused with mutualism. One or both species involved in the interaction may be obligate, meaning they cannot survive in the short or long term without the other species. Though mutualism has historically received less attention than other interactions such as predation, it is an important subject in ecology. Examples include cleaning symbiosis, gut flora, Müllerian mimicry, and nitrogen fixation by bacteria in the root nodules of legumes.

Commensalism

Commensalism benefits one organism and the other organism is neither benefited nor harmed. It occurs when one organism takes benefits by interacting with another organism by which the host organism is not affected. A good example is a remora living with a shark. Remoras eat leftover food from the shark. The shark is not affected in the process, as remoras eat only leftover food of the shark, which does not deplete the shark's resources.

Parasitism

Parasitism is a relationship between species, where one organism, the parasite, lives on or in another organism, the host, causing it some harm, and is adapted structurally to this way of life. The parasite either feeds on the host, or, in the case of intestinal parasites, consumes some of its food.

Neutralism

Neutralism (a term introduced by Eugene Odum) describes the relationship between two species that interact but do not affect each other. Examples of true neutralism are virtually impossible to prove; the term is in practice used to describe situations where interactions are negligible or insignificant.

Amensalism

Amensalism (a term introduced by Haskell) is an interaction where an organism inflicts harm to another organism without any costs or benefits received by itself. A clear case of amensalism is where sheep or cattle trample grass. Whilst the presence of the grass causes negligible detrimental effects to the animal's hoof, the grass suffers from being crushed. Amensalism is often used to describe strongly asymmetrical competitive interactions, such as has been observed between the Spanish ibex and weevils of the genus Timarcha which feed upon the same type of shrub. Whilst the presence of the weevil has almost no influence on food availability, the presence of ibex has an enormous detrimental effect on weevil numbers, as they consume significant quantities of plant matter and incidentally ingest the weevils upon it.

Competition

Male-male interference competition in red deer.

Competition can be defined as an interaction between organisms or species, in which the fitness of one is lowered by the presence of another. Competition is often for a resource such as food, water, or territory in limited supply, or for access to females for reproduction. Competition among members of the same species is known as intraspecific competition, while competition between individuals of different species is known as interspecific competition. According to the competitive exclusion principle, species less suited to compete for resources should either adapt or die out. According to evolutionary theory, this competition within and between species for resources plays a critical role in natural selection.

References

- Markov, a; korotayev, a (2007). "Phanerozoic marine biodiversity follows a hyperbolic trend". Palaeoworld. 16 (4): 311–318. Doi:10.1016/J.Palwor.2007.01.002.

- Define-biodiversity, biodiversity: greenfacts.Org, retrieved 20 july, 2019

- Four-types-biodiversity: sciencing.Com, retrieved 15 may, 2019

- Tuomisto, h. 2010. A consistent terminology for quantifying species diversity? Yes, it does exist. Oecologia 4: 853–860. Doi:10.1007/S00442-010-1812-0

- "National biological information infrastructure". Introduction to genetic diversity. U.S. Geological survey. Archived from the originalon february 25, 2011. Retrieved march 1, 2011.

- Measuring-biodiversity, biodiversity, ecosystems: enviroliteracy.Org, retrieved 12 april 2019

- Wright, alan f. (September 2005). Genetic variation: polymorphisms and mutations. Encyclopedia of life sciences. Doi:10.1038/Npg.Els.0005005. Isbn 978-0470016176.

- Define-biodiversity, biodiversity: greenfacts.Org, retrieved 16 june, 2019

- Gaston, kevin j. (11 May 2000). "Global patterns in biodiversity". Nature. 405 (6783): 220–227. Doi:10.1038/35012228. Pmid 10821282.

- Geographic-distribution-biodiversity, biodiversity: healingearth.Ijep.Net, retrieved 10 july, 2019

- Bar-yam. "Predator-prey relationships". New england complex systems institute. Retrieved 7 september 2018.

Evolution: Types and Processes

The change in heritable characteristics of the biological populations is referred to as evolution. It is broadly divided into a few categories such as divergent evolution, convergent evolution, macroevolution and microevolution. All these diverse processes and types of evolution have been carefully analyzed in this chapter.

EVOLUTION AND BIODIVERSITY

Biodiversity has a key role in maintaining healthy ecosystems and thereby sustaining ecosystem services to the ever-growing human population. To get an idea of the range of ecosystem services that we use daily, think of how much energy and time it would cost to make Mars hospitable for human life, for example, in terms of atmosphere regulation, freshwater production, soil formation, nutrient cycles, regulation of climate, etc. On our own planet, that process took four billion years and required the contribution of a vast amount of functions performed by different life forms, ultimately driven by evolution and that is only the top of the iceberg.

Unfortunately, the ecosystems that we so exploit and dearly need for our long-term survival and welfare are jeopardized by our own actions. Global change, triggered by human activities, is all around us. The pervasive effects of climate change, habitat loss and fragmentation, overharvesting, pollution, altered nutrient cycling, invasive species and interactions thereof affect virtually all Earth's ecosystems. With seven billion people consuming natural resources more rapidly than they are created, we are at the onset of a major environmental revolution. As a consequence, species are already shifting, expanding, disappearing, changing their behaviour and phenology, exploiting newly available food resources and abandoning scarcer ones. Ecosystems are changing too, driven by changes in environmental drivers and by the reshuffling of their biota into previously unknown combinations of species. The interplay of all these processes makes the forecast of changes in ecosystem services a daunting task.

All these changes are likely to have a bearing on and be influenced by the evolutionary forces at play. The main legacy of Charles Darwin was to make us realize that we owe everything, including the formation of our own species, to evolution. We thus learned that the history of life is driven by evolution. But what about the future? What is the contribution of evolution to these ecological changes? And, probably most relevant to policy: what is the potential of evolutionary processes to exacerbate or mitigate the effects of global change? Is evolutionary biology just a scientific gimmick compared to the real problems that we are facing?

Until a decade or so ago, evolutionary change was broadly assumed to happen on a vastly longer time scale than ecological change. As a corollary, our view on biodiversity and ecosystem functioning has often been static, trying to conserve biodiversity as it is, and preferably, as it once was. Just like our ecosystems, however, this paradigm is shifting. The closer we look at adaptive evolution, often with the aid of new biological insights and technological advances, the faster it seems to happen. Evolution and ecology are proving to be so heavily entwined that the distinction is becoming increasingly hard to make. This knowledge profoundly affects our thinking on how evolution affects patterns of biodiversity, especially in the face of global change. Adaptive responses to climate change, for example, have been shown to occur within a single generation. Contemporary evolution is probably more important than we assumed to date and is, therefore, likely to mediate the response of populations, species, communities and ecosystems to both gradual and sudden environmental change.

In April 2010, the European Platform for Biodiversity Research Strategy hosted the meeting 'Evolution and Biodiversity: The evolutionary basis of biodiversity and its potential for adaptation to global change', funded by the BioStrat project. The meeting was preceded by an electronic conference that lasted 21 days and gathered over 62 contributors and more than 1600 participants. Both the conference and the meeting revolved around three main themes: the evolutionary basis of biodiversity, evolutionary responses to global change and evolution in complex systems and co-evolutionary networks. This special issue builds from the diverse arrange of contributions made at the conference and aims to provide a diverse and interdisciplinary perspective on the interplay between evolutionary and ecological responses in the face of global change.

Biodiversity is the result of 3.5 billion years of evolution. The origin of life has not been definitely established by science, however some evidence suggests that life may already have been well-established only a few hundred million years after the formation of the Earth. Until approximately 600 million years ago, all life consisted of microorganisms – archaea, bacteria, and single-celled protozoans and protists.

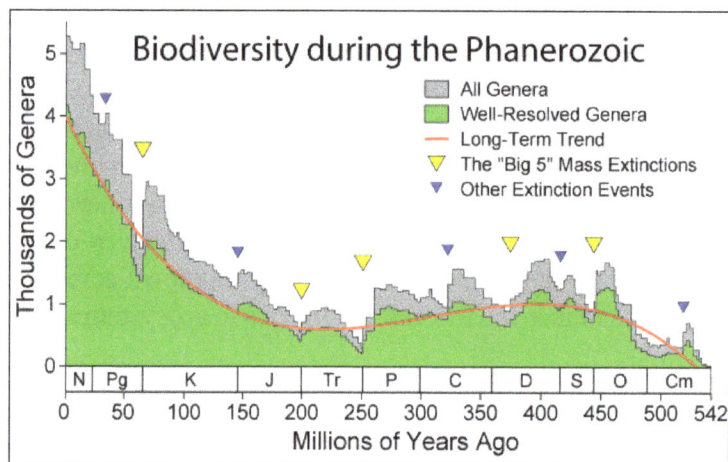

Apparent marine fossil diversity during the Phanerozoic.

The history of biodiversity during the Phanerozoic (the last 540 million years), starts with rapid growth during the Cambrian explosion—a period during which nearly every phylum of multicellular organisms first appeared. Over the next 400 million years or so, invertebrate diversity showed little overall trend and vertebrate diversity shows an overall exponential trend. This dramatic rise in diversity was marked by periodic, massive losses of diversity classified as mass extinction events. A significant loss occurred when rainforests collapsed in the carboniferous. The worst was the Permian-Triassic extinction event, 251 million years ago. Vertebrates took 30 million years to recover from this event.

The fossil record suggests that the last few million years featured the greatest biodiversity in history. However, not all scientists support this view, since there is uncertainty as to how strongly the fossil record is biased by the greater availability and preservation of recent geologic sections. Some scientists believe that corrected for sampling artifacts, modern biodiversity may not be much different from biodiversity 300 million years ago, whereas others consider the fossil record reasonably reflective of the diversification of life. Estimates of the present global macroscopic species diversity vary from 2 million to 100 million, with a best estimate of somewhere near 9 million, the vast majority arthropods. Diversity appears to increase continually in the absence of natural selection.

Diversification

The existence of a global carrying capacity, limiting the amount of life that can live at once, is debated, as is the question of whether such a limit would also cap the number of species. While records of life in the sea shows a logistic pattern of growth, life on land (insects, plants and tetrapods) shows an exponential rise in diversity. As one author states, "Tetrapods have not yet invaded 64 per cent of potentially habitable modes and it could be that without human influence the ecological and taxonomic diversity of tetrapods would continue to increase in an exponential fashion until most or all of the available ecospace is filled."

It also appears that the diversity continue to increase over time, especially after mass extinctions.

On the other hand, changes through the Phanerozoic correlate much better with the hyperbolic model (widely used in population biology, demography and macrosociology, as well as fossil biodiversity) than with exponential and logistic models. The latter models imply that changes in diversity are guided by a first-order positive feedback (more ancestors, more descendants) and/ or a negative feedback arising from resource limitation. Hyperbolic model implies a second-order positive feedback. The hyperbolic pattern of the world population growth arises from a second-order positive feedback between the population size and the rate of technological growth. The hyperbolic character of biodiversity growth can be similarly accounted for by a feedback between diversity and community structure complexity. The similarity between the curves of biodiversity and human population probably comes from the fact that both are derived from the interference of the hyperbolic trend with cyclical and stochastic dynamics.

Most biologists agree however that the period since human emergence is part of a new mass extinction, named the Holocene extinction event, caused primarily by the impact humans are having on the environment. It has been argued that the present rate of extinction is sufficient to eliminate most species on the planet Earth within 100 years.

In 2011, in his Biodiversity-related Niches Differentiation Theory, Roberto Cazzolla Gatti proposed that species themselves are the architects of biodiversity, by proportionally increasing the number of potentially available niches in a given ecosystem. This study led to the idea that biodiversity is autocatalytic. An ecosystem of interdependent species can be, therefore, considered as an emergent autocatalytic set (a self-sustaining network of mutually "catalytic" entities), where one (group of) species enables the existence of (i.e., creates niches for) other species. This view offers a possible answer to the fundamental question of why so many species can coexist in the same ecosystem.

New species are regularly discovered (on average between 5–10,000 new species each year, most of them insects) and many, though discovered, are not yet classified (estimates are that nearly 90% of all arthropods are not yet classified). Most of the terrestrial diversity is found in tropical forests and in general, land has more species than the ocean; some 8.7 million species may exists on Earth, of which some 2.1 million live in the ocean.

Evolution is change in the heritable characteristics of biological populations over successive generations. These characteristics are the expressions of genes that are passed on from parent to offspring during reproduction. Different characteristics tend to exist within any given population as a result of mutation, genetic recombination and other sources of genetic variation. Evolution occurs when evolutionary processes such as natural selection (including sexual selection) and genetic drift act on this variation, resulting in certain characteristics becoming more common or rare within a population. It is this process of evolution that has given rise to biodiversity at every level of biological organisation, including the levels of species, individual organisms and molecules.

The scientific theory of evolution by natural selection was proposed by Charles Darwin and Alfred Russel Wallace in the mid-19th century and was set out in detail in Darwin's book On the Origin of Species (1859). Evolution by natural selection was first demonstrated by the observation that more offspring are often produced than can possibly survive. This is followed by three observable facts about living organisms: 1) traits vary among individuals with respect to their morphology, physiology and behaviour (phenotypic variation), 2) different traits confer different rates of survival and reproduction (differential fitness) and 3) traits can be passed from generation to generation (heritability of fitness). Thus, in successive generations members of a population are more likely to be replaced by the progenies of parents with favourable characteristics that have enabled them to survive and reproduce in their respective environments. In the early 20th century, other competing ideas of evolution such as mutationism and orthogenesis were refuted as the modern synthesis reconciled Darwinian evolution with classical genetics, which established adaptive evolution as being caused by natural selection acting on Mendelian genetic variation.

All life on Earth shares a last universal common ancestor (LUCA) that lived approximately 3.5–3.8 billion years ago. The fossil record includes a progression from early biogenic graphite, to microbial mat fossils, to fossilised multicellular organisms. Existing patterns of biodiversity have been shaped by repeated formations of new species (speciation), changes within species (anagenesis) and loss of species (extinction) throughout the evolutionary history of life on Earth. Morphological and biochemical traits are more similar among species that share a more recent common ancestor, and can be used to reconstruct phylogenetic trees.

Evolutionary biologists have continued to study various aspects of evolution by forming and testing hypotheses as well as constructing theories based on evidence from the field or laboratory and on data generated by the methods of mathematical and theoretical biology. Their discoveries have influenced not just the development of biology but numerous other scientific and industrial fields, including agriculture, medicine and computer science.

Heredity

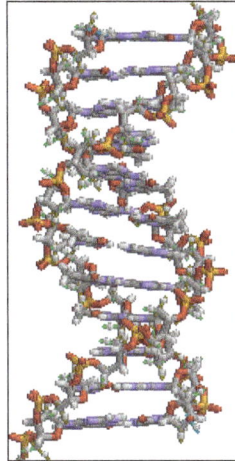

DNA structure. Bases are in the centre, surrounded
by phosphate–sugar chains in a double helix.

Evolution in organisms occurs through changes in heritable traits—the inherited characteristics of an organism. In humans, for example, eye colour is an inherited characteristic and an individual might inherit the "brown-eye trait" from one of their parents. Inherited traits are controlled by genes and the complete set of genes within an organism's genome (genetic material) is called its genotype.

The complete set of observable traits that make up the structure and behaviour of an organism is called its phenotype. These traits come from the interaction of its genotype with the environment. As a result, many aspects of an organism's phenotype are not inherited. For example, suntanned skin comes from the interaction between a person's genotype and sunlight; thus, suntans are not passed on to people's children. However, some people tan more easily than others, due to differences in genotypic variation; a striking example are people with the inherited trait of albinism, who do not tan at all and are very sensitive to sunburn.

Heritable traits are passed from one generation to the next via DNA, a molecule that encodes genetic information. DNA is a long biopolymer composed of four types of bases. The sequence of bases along a particular DNA molecule specify the genetic information, in a manner similar to a sequence of letters spelling out a sentence. Before a cell divides, the DNA is copied, so that each of the resulting two cells will inherit the DNA sequence. Portions of a DNA molecule that specify a single functional unit are called genes; different genes have different sequences of bases. Within cells, the long strands of DNA form condensed structures called chromosomes. The specific location of a DNA sequence within a chromosome is known as a locus. If the DNA sequence at a locus varies between individuals, the different forms of this sequence are called alleles. DNA sequences can change through mutations, producing new alleles. If a mutation occurs within a gene, the new allele may affect the trait that the gene controls, altering the phenotype of the organism. However, while this simple correspondence between an allele and a trait works in some cases, most traits are more complex and are controlled by quantitative trait loci (multiple interacting genes).

Recent findings have confirmed important examples of heritable changes that cannot be explained by changes to the sequence of nucleotides in the DNA. These phenomena are classed as epigenetic inheritance systems. DNA methylation marking chromatin, self-sustaining metabolic loops, gene silencing by RNA interference and the three-dimensional conformation of proteins (such as prions) are areas where epigenetic inheritance systems have been discovered at the organismic level. Developmental biologists suggest that complex interactions in genetic networks and communication among cells can lead to heritable variations that may underlay some of the mechanics in developmental plasticity and canalisation. Heritability may also occur at even larger scales. For example, ecological inheritance through the process of niche construction is defined by the regular and repeated activities of organisms in their environment. This generates a legacy of effects that modify and feed back into the selection regime of subsequent generations. Descendants inherit genes plus environmental characteristics generated by the ecological actions of ancestors. Other examples of heritability in evolution that are not under the direct control of genes include the inheritance of cultural traits and symbiogenesis.

Variation

White peppered moth

Black morph in peppered moth evolution

An individual organism's phenotype results from both its genotype and the influence from the environment it has lived in. A substantial part of the phenotypic variation in a population is caused by genotypic variation. The modern evolutionary synthesis defines evolution as the change over time in this genetic variation. The frequency of one particular allele will become more or less prevalent relative to other forms of that gene. Variation disappears when a new allele reaches the point of fixation—when it either disappears from the population or replaces the ancestral allele entirely.

Natural selection will only cause evolution if there is enough genetic variation in a population. Before the discovery of Mendelian genetics, one common hypothesis was blending inheritance. But with blending inheritance, genetic variance would be rapidly lost, making evolution by natural selection implausible. The Hardy–Weinberg principle provides the solution to how variation is maintained in a population with Mendelian inheritance. The frequencies of alleles (variations in a gene) will remain constant in the absence of selection, mutation, migration and genetic drift.

Variation comes from mutations in the genome, reshuffling of genes through sexual reproduction and migration between populations (gene flow). Despite the constant introduction of new variation through mutation and gene flow, most of the genome of a species is identical in all individuals of that species. However, even relatively small differences in genotype can lead to dramatic differences in phenotype: for example, chimpanzees and humans differ in only about 5% of their genomes.

Mutation

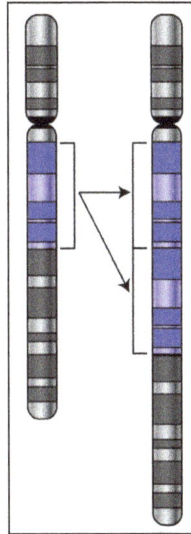

Duplication of part of a chromosome.

Mutations are changes in the DNA sequence of a cell's genome. When mutations occur, they may alter the product of a gene, or prevent the gene from functioning, or have no effect. Based on studies in the fly Drosophila melanogaster, it has been suggested that if a mutation changes a protein produced by a gene, this will probably be harmful, with about 70% of these mutations having damaging effects, and the remainder being either neutral or weakly beneficial.

Mutations can involve large sections of a chromosome becoming duplicated (usually by genetic recombination), which can introduce extra copies of a gene into a genome. Extra copies of genes are a major source of the raw material needed for new genes to evolve. This is important because most new genes evolve within gene families from pre-existing genes that share common ancestors. For example, the human eye uses four genes to make structures that sense light: three for colour vision and one for night vision; all four are descended from a single ancestral gene.

New genes can be generated from an ancestral gene when a duplicate copy mutates and acquires a new function. This process is easier once a gene has been duplicated because it increases the redundancy of the system; one gene in the pair can acquire a new function while the other copy continues to perform its original function. Other types of mutations can even generate entirely new genes from previously noncoding DNA.

The generation of new genes can also involve small parts of several genes being duplicated, with these fragments then recombining to form new combinations with new functions. When new genes are assembled from shuffling pre-existing parts, domains act as modules with simple independent functions, which can be mixed together to produce new combinations with new and complex functions. For example, polyketide synthases are large enzymes that make antibiotics; they contain up to one hundred independent domains that each catalyse one step in the overall process, like a step in an assembly line.

Sex and Recombination

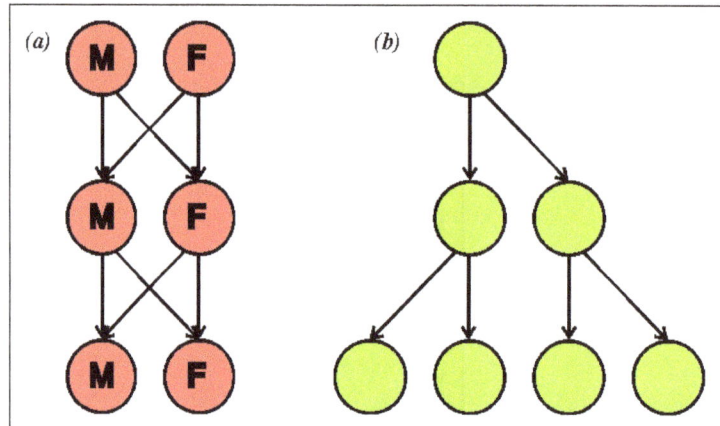

This diagram illustrates the twofold cost of sex. If each individual were to contribute to the same number of offspring (two), (a) the sexual population remains the same size each generation, where the (b) Asexual reproduction population doubles in size each generation.

In asexual organisms, genes are inherited together, or linked, as they cannot mix with genes of other organisms during reproduction. In contrast, the offspring of sexual organisms contain random mixtures of their parents' chromosomes that are produced through independent assortment. In a related process called homologous recombination, sexual organisms exchange DNA between two matching chromosomes. Recombination and reassortment do not alter allele frequencies, but instead change which alleles are associated with each other, producing offspring with new combinations of alleles. Sex usually increases genetic variation and may increase the rate of evolution.

The two-fold cost of sex was first described by John Maynard Smith. The first cost is that in sexually dimorphic species only one of the two sexes can bear young. (This cost does not apply to hermaphroditic species, like most plants and many invertebrates.) The second cost is that any individual who reproduces sexually can only pass on 50% of its genes to any individual offspring, with even less passed on as each new generation passes. Yet sexual reproduction is the more common means of reproduction among eukaryotes and multicellular organisms. The Red Queen hypothesis has been used to explain the significance of sexual reproduction as a means to enable continual evolution and adaptation in response to coevolution with other species in an ever-changing environment.

Gene Flow

Gene flow is the exchange of genes between populations and between species. It can therefore be a source of variation that is new to a population or to a species. Gene flow can be caused by the movement of individuals between separate populations of organisms, as might be caused by the movement of mice between inland and coastal populations, or the movement of pollen between heavy-metal-tolerant and heavy-metal-sensitive populations of grasses.

Gene transfer between species includes the formation of hybrid organisms and horizontal gene transfer. Horizontal gene transfer is the transfer of genetic material from one organism to another organism that is not its offspring; this is most common among bacteria. In medicine, this contributes to the spread of antibiotic resistance, as when one bacteria acquires resistance genes it can rapidly transfer them to other species. Horizontal transfer of genes from bacteria to eukaryotes such as the yeast Saccharomyces cerevisiae and the adzuki bean weevil Callosobruchus chinensis has occurred. An example of larger-scale transfers are the eukaryotic bdelloid rotifers, which have received a range of genes from bacteria, fungi and plants. Viruses can also carry DNA between organisms, allowing transfer of genes even across biological domains.

Large-scale gene transfer has also occurred between the ancestors of eukaryotic cells and bacteria, during the acquisition of chloroplasts and mitochondria. It is possible that eukaryotes themselves originated from horizontal gene transfers between bacteria and archaea.

Mechanisms

From a neo-Darwinian perspective, evolution occurs when there are changes in the frequencies of alleles within a population of interbreeding organisms, for example, the allele for black colour in a population of moths becoming more common. Mechanisms that can lead to changes in allele frequencies include natural selection, genetic drift, genetic hitchhiking, mutation and gene flow.

Natural Selection

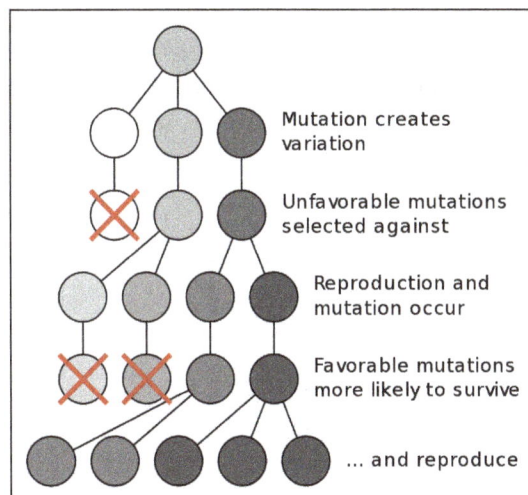

Mutation followed by natural selection results in a population with darker colouration.

Evolution by means of natural selection is the process by which traits that enhance survival and reproduction become more common in successive generations of a population. It has often been called a "self-evident" mechanism because it necessarily follows from three simple facts:

- Variation exists within populations of organisms with respect to morphology, physiology, and behaviour (phenotypic variation).

- Different traits confer different rates of survival and reproduction (differential fitness).

- These traits can be passed from generation to generation (heritability of fitness).

More offspring are produced than can possibly survive, and these conditions produce competition between organisms for survival and reproduction. Consequently, organisms with traits that give them an advantage over their competitors are more likely to pass on their traits to the next generation than those with traits that do not confer an advantage. This teleonomy is the quality whereby the process of natural selection creates and preserves traits that are seemingly fitted for the functional roles they perform. Consequences of selection include nonrandom mating and genetic hitchhiking.

The central concept of natural selection is the evolutionary fitness of an organism. Fitness is measured by an organism's ability to survive and reproduce, which determines the size of its genetic contribution to the next generation. However, fitness is not the same as the total number of offspring: instead fitness is indicated by the proportion of subsequent generations that carry an organism's genes. For example, if an organism could survive well and reproduce rapidly, but its offspring were all too small and weak to survive, this organism would make little genetic contribution to future generations and would thus have low fitness.

If an allele increases fitness more than the other alleles of that gene, then with each generation this allele will become more common within the population. These traits are said to be "selected for." Examples of traits that can increase fitness are enhanced survival and increased fecundity. Conversely, the lower fitness caused by having a less beneficial or deleterious allele results in this allele becoming rarer—they are "selected against." Importantly, the fitness of an allele is not a fixed characteristic; if the environment changes, previously neutral or harmful traits may become beneficial and previously beneficial traits become harmful. However, even if the direction of selection does reverse in this way, traits that were lost in the past may not re-evolve in an identical form. However, a re-activation of dormant genes, as long as they have not been eliminated from the genome and were only suppressed perhaps for hundreds of generations, can lead to the re-occurrence of traits thought to be lost like hindlegs in dolphins, teeth in chickens, wings in wingless stick insects, tails and additional nipples in humans etc. "Throwbacks" such as these are known as atavisms.

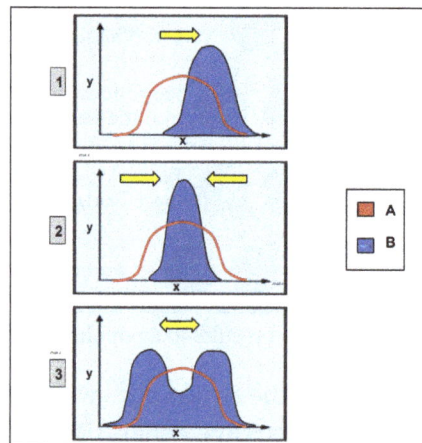

These charts depict the different types of genetic selection. On each graph, the x-axis variable is the type of phenotypic trait and the y-axis variable is the number of organisms. Group A is the original population and Group B is the population after selection. Graph 1 shows directional selection, in which a single extreme phenotype is favoured. Graph 2 depicts stabilizing selection, where the intermediate phenotype is favoured over the extreme traits. Graph 3 shows disruptive selection, in which the extreme phenotypes are favoured over the intermediate.

Natural selection within a population for a trait that can vary across a range of values, such as height, can be categorised into three different types. The first is directional selection, which is a shift in the average value of a trait over time—for example, organisms slowly getting taller. Secondly, disruptive selection is selection for extreme trait values and often results in two different values becoming most common, with selection against the average value. This would be when either short or tall organisms had an advantage, but not those of medium height. Finally, in stabilising selection there is selection against extreme trait values on both ends, which causes a decrease in variance around the average value and less diversity. This would, for example, cause organisms to eventually have a similar height.

A special case of natural selection is sexual selection, which is selection for any trait that increases mating success by increasing the attractiveness of an organism to potential mates. Traits that evolved through sexual selection are particularly prominent among males of several animal species. Although sexually favoured, traits such as cumbersome antlers, mating calls, large body size and bright colours often attract predation, which compromises the survival of individual males. This survival disadvantage is balanced by higher reproductive success in males that show these hard-to-fake, sexually selected traits.

Natural selection most generally makes nature the measure against which individuals and individual traits, are more or less likely to survive. "Nature" in this sense refers to an ecosystem, that is, a system in which organisms interact with every other element, physical as well as biological, in their local environment. Eugene Odum, a founder of ecology, defined an ecosystem as: "Any unit that includes all of the organisms in a given area interacting with the physical environment so that a flow of energy leads to clearly defined trophic structure, biotic diversity, and material cycles (i.e., exchange of materials between living and nonliving parts) within the system...." Each population within an ecosystem occupies a distinct niche, or position, with distinct relationships to other parts of the system. These relationships involve the life history of the organism, its position in the food chain and its geographic range. This broad understanding of nature enables scientists to delineate specific forces which, together, comprise natural selection.

Natural selection can act at different levels of organisation, such as genes, cells, individual organisms, groups of organisms and species. Selection can act at multiple levels simultaneously. An example of selection occurring below the level of the individual organism are genes called transposons, which can replicate and spread throughout a genome. Selection at a level above the individual, such as group selection, may allow the evolution of cooperation.

Biased Mutation

In addition to being a major source of variation, mutation may also function as a mechanism of evolution when there are different probabilities at the molecular level for different mutations to occur, a process known as mutation bias. If two genotypes, for example one with the nucleotide G and another with the nucleotide A in the same position, have the same fitness, but mutation from G to A happens more often than mutation from A to G, then genotypes with A will tend to evolve. Different insertion vs. deletion mutation biases in different taxa can lead to the evolution of different genome sizes. Developmental or mutational biases have also been observed in morphological evolution. For example, according to the phenotype-first theory of evolution, mutations can eventually cause the genetic assimilation of traits that were previously induced by the environment.

Mutation bias effects are superimposed on other processes. If selection would favour either one out of two mutations, but there is no extra advantage to having both, then the mutation that occurs the most frequently is the one that is most likely to become fixed in a population. Mutations leading to the loss of function of a gene are much more common than mutations that produce a new, fully functional gene. Most loss of function mutations are selected against. But when selection is weak, mutation bias towards loss of function can affect evolution. For example, pigments are no longer useful when animals live in the darkness of caves, and tend to be lost. This kind of loss of function can occur because of mutation bias, and/or because the function had a cost, and once the benefit of the function disappeared, natural selection leads to the loss. Loss of sporulation ability in Bacillus subtilis during laboratory evolution appears to have been caused by mutation bias, rather than natural selection against the cost of maintaining sporulation ability. When there is no selection for loss of function, the speed at which loss evolves depends more on the mutation rate than it does on the effective population size, indicating that it is driven more by mutation bias than by genetic drift. In parasitic organisms, mutation bias leads to selection pressures as seen in Ehrlichia. Mutations are biased towards antigenic variants in outer-membrane proteins.

Genetic Drift

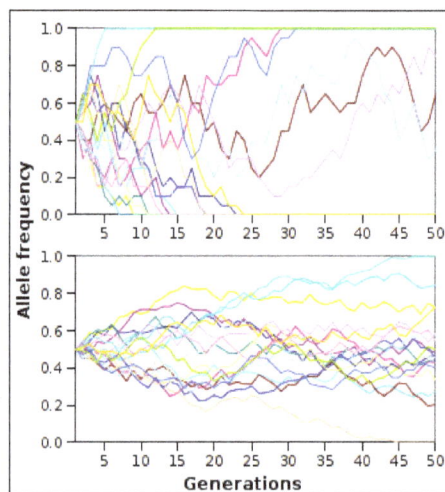

Simulation of genetic drift of 20 unlinked alleles in populations of 10 (top) and 100 (bottom). Drift to fixation is more rapid in the smaller population.

Genetic drift is the random fluctuations of allele frequencies within a population from one generation to the next. When selective forces are absent or relatively weak, allele frequencies are equally likely to drift upward or downward at each successive generation because the alleles are subject to sampling error. This drift halts when an allele eventually becomes fixed, either by disappearing from the population or replacing the other alleles entirely. Genetic drift may therefore eliminate some alleles from a population due to chance alone. Even in the absence of selective forces, genetic drift can cause two separate populations that began with the same genetic structure to drift apart into two divergent populations with different sets of alleles.

The neutral theory of molecular evolution proposed that most evolutionary changes are the result of the fixation of neutral mutations by genetic drift. Hence, in this model, most genetic changes in a population are the result of constant mutation pressure and genetic drift. This form of the neutral theory is now largely abandoned, since it does not seem to fit the genetic variation seen in nature.

However, a more recent and better-supported version of this model is the nearly neutral theory, where a mutation that would be effectively neutral in a small population is not necessarily neutral in a large population. Other alternative theories propose that genetic drift is dwarfed by other stochastic forces in evolution, such as genetic hitchhiking, also known as genetic draft.

The time for a neutral allele to become fixed by genetic drift depends on population size, with fixation occurring more rapidly in smaller populations. The number of individuals in a population is not critical, but instead a measure known as the effective population size. The effective population is usually smaller than the total population since it takes into account factors such as the level of inbreeding and the stage of the lifecycle in which the population is the smallest. The effective population size may not be the same for every gene in the same population.

It is usually difficult to measure the relative importance of selection and neutral processes, including drift. The comparative importance of adaptive and non-adaptive forces in driving evolutionary change is an area of current research.

Genetic Hitchhiking

Recombination allows alleles on the same strand of DNA to become separated. However, the rate of recombination is low (approximately two events per chromosome per generation). As a result, genes close together on a chromosome may not always be shuffled away from each other and genes that are close together tend to be inherited together, a phenomenon known as linkage. This tendency is measured by finding how often two alleles occur together on a single chromosome compared to expectations, which is called their linkage disequilibrium. A set of alleles that is usually inherited in a group is called a haplotype. This can be important when one allele in a particular haplotype is strongly beneficial: natural selection can drive a selective sweep that will also cause the other alleles in the haplotype to become more common in the population; this effect is called genetic hitchhiking or genetic draft. Genetic draft caused by the fact that some neutral genes are genetically linked to others that are under selection can be partially captured by an appropriate effective population size.

Gene Flow

Gene flow involves the exchange of genes between populations and between species. The presence or absence of gene flow fundamentally changes the course of evolution. Due to the complexity of organisms, any two completely isolated populations will eventually evolve genetic incompatibilities through neutral processes, as in the Bateson-Dobzhansky-Muller model, even if both populations remain essentially identical in terms of their adaptation to the environment.

If genetic differentiation between populations develops, gene flow between populations can introduce traits or alleles which are disadvantageous in the local population and this may lead to organisms within these populations evolving mechanisms that prevent mating with genetically distant populations, eventually resulting in the appearance of new species. Thus, exchange of genetic information between individuals is fundamentally important for the development of the Biological Species Concept (BSC).

During the development of the modern synthesis, Sewall Wright developed his shifting balance theory, which regarded gene flow between partially isolated populations as an important aspect of adaptive evolution. However, recently there has been substantial criticism of the importance of the shifting balance theory.

Outcomes

Evolution influences every aspect of the form and behaviour of organisms. Most prominent are the specific behavioural and physical adaptations that are the outcome of natural selection. These adaptations increase fitness by aiding activities such as finding food, avoiding predators or attracting mates. Organisms can also respond to selection by cooperating with each other, usually by aiding their relatives or engaging in mutually beneficial symbiosis. In the longer term, evolution produces new species through splitting ancestral populations of organisms into new groups that cannot or will not interbreed.

These outcomes of evolution are distinguished based on time scale as macroevolution versus microevolution. Macroevolution refers to evolution that occurs at or above the level of species, in particular speciation and extinction; whereas microevolution refers to smaller evolutionary changes within a species or population, in particular shifts in allele frequency and adaptation. In general, macroevolution is regarded as the outcome of long periods of microevolution. Thus, the distinction between micro- and macroevolution is not a fundamental one—the difference is simply the time involved. However, in macroevolution, the traits of the entire species may be important. For instance, a large amount of variation among individuals allows a species to rapidly adapt to new habitats, lessening the chance of it going extinct, while a wide geographic range increases the chance of speciation, by making it more likely that part of the population will become isolated. In this sense, microevolution and macroevolution might involve selection at different levels—with microevolution acting on genes and organisms, versus macroevolutionary processes such as species selection acting on entire species and affecting their rates of speciation and extinction.

A common misconception is that evolution has goals, long-term plans, or an innate tendency for "progress", as expressed in beliefs such as orthogenesis and evolutionism; realistically however, evolution has no long-term goal and does not necessarily produce greater complexity. Although complex species have evolved, they occur as a side effect of the overall number of organisms increasing and simple forms of life still remain more common in the biosphere. For example, the overwhelming majority of species are microscopic prokaryotes, which form about half the world's biomass despite their small size, and constitute the vast majority of Earth's biodiversity. Simple organisms have therefore been the dominant form of life on Earth throughout its history and continue to be the main form of life up to the present day, with complex life only appearing more diverse because it is more noticeable. Indeed, the evolution of microorganisms is particularly important to modern evolutionary research, since their rapid reproduction allows the study of experimental evolution and the observation of evolution and adaptation in real time.

Adaptation

Adaptation is the process that makes organisms better suited to their habitat. Also, the term adaptation may refer to a trait that is important for an organism's survival. For example, the adaptation of horses' teeth to the grinding of grass. By using the term adaptation for the evolutionary process and adaptive trait for the product (the bodily part or function), the two senses of the word may be distinguished. Adaptations are produced by natural selection. The following definitions are due to Theodosius Dobzhansky:

- Adaptation is the evolutionary process whereby an organism becomes better able to live in its habitat or habitats.

- Adaptedness is the state of being adapted: the degree to which an organism is able to live and reproduce in a given set of habitats.

- An adaptive trait is an aspect of the developmental pattern of the organism which enables or enhances the probability of that organism surviving and reproducing.

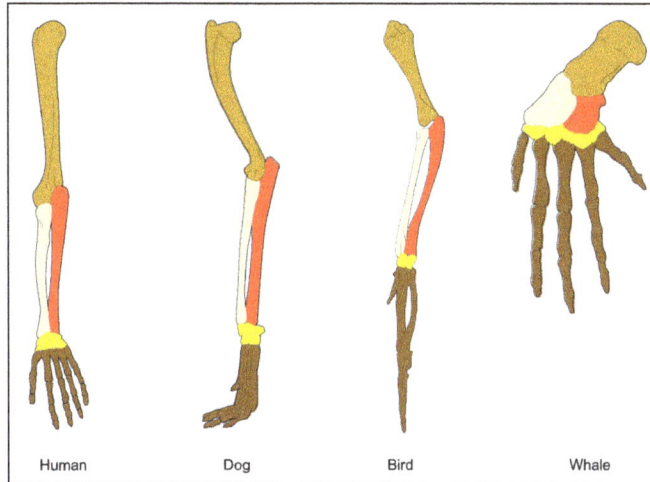

| Human | Dog | Bird | Whale |

Homologous bones in the limbs of tetrapods. The bones of these animals have the same basic structure,but have been adapted for specific uses.

Adaptation may cause either the gain of a new feature, or the loss of an ancestral feature. An example that shows both types of change is bacterial adaptation to antibiotic selection, with genetic changes causing antibiotic resistance by both modifying the target of the drug, or increasing the activity of transporters that pump the drug out of the cell. Other striking examples are the bacteria Escherichia coli evolving the ability to use citric acid as a nutrient in a long-term laboratory experiment, Flavobacterium evolving a novel enzyme that allows these bacteria to grow on the by-products of nylon manufacturing, and the soil bacterium Sphingobium evolving an entirely new metabolic pathway that degrades the synthetic pesticide pentachlorophenol. An interesting but still controversial idea is that some adaptations might increase the ability of organisms to generate genetic diversity and adapt by natural selection (increasing organisms' evolvability).

Adaptation occurs through the gradual modification of existing structures. Consequently, structures with similar internal organisation may have different functions in related organisms. This is the result of a single ancestral structure being adapted to function in different ways. The bones within bat wings, for example, are very similar to those in mice feet and primate hands, due to the descent of all these structures from a common mammalian ancestor. However, since all living organisms are related to some extent, even organs that appear to have little or no structural similarity, such as arthropod, squid and vertebrate eyes, or the limbs and wings of arthropods and vertebrates, can depend on a common set of homologous genes that control their assembly and function; this is called deep homology.

During evolution, some structures may lose their original function and become vestigial structures. Such structures may have little or no function in a current species, yet have a clear function in ancestral species, or other closely related species. Examples include pseudogenes, the non-functional remains of eyes in blind cave-dwelling fish, wings in flightless birds, the presence of hip bones in whales and snakes, and sexual traits in organisms that reproduce via asexual reproduction.

Examples of vestigial structures in humans include wisdom teeth, the coccyx, the vermiform appendix, and other behavioural vestiges such as goose bumps and primitive reflexes.

However, many traits that appear to be simple adaptations are in fact exaptations: structures originally adapted for one function, but which coincidentally became somewhat useful for some other function in the process. One example is the African lizard Holaspis guentheri, which developed an extremely flat head for hiding in crevices, as can be seen by looking at its near relatives. However, in this species, the head has become so flattened that it assists in gliding from tree to tree—an exaptation. Within cells, molecular machines such as the bacterial flagella and protein sorting machinery evolved by the recruitment of several pre-existing proteins that previously had different functions. Another example is the recruitment of enzymes from glycolysis and xenobiotic metabolism to serve as structural proteins called crystallins within the lenses of organisms' eyes.

An area of current investigation in evolutionary developmental biology is the developmental basis of adaptations and exaptations. This research addresses the origin and evolution of embryonic development and how modifications of development and developmental processes produce novel features. These studies have shown that evolution can alter development to produce new structures, such as embryonic bone structures that develop into the jaw in other animals instead forming part of the middle ear in mammals. It is also possible for structures that have been lost in evolution to reappear due to changes in developmental genes, such as a mutation in chickens causing embryos to grow teeth similar to those of crocodiles. It is now becoming clear that most alterations in the form of organisms are due to changes in a small set of conserved genes.

Coevolution

Interactions between organisms can produce both conflict and cooperation. When the interaction is between pairs of species, such as a pathogen and a host, or a predator and its prey, these species can develop matched sets of adaptations. Here, the evolution of one species causes adaptations in a second species. These changes in the second species then, in turn, cause new adaptations in the first species. This cycle of selection and response is called coevolution. An example is the production of tetrodotoxin in the rough-skinned newt and the evolution of tetrodotoxin resistance in its predator, the common garter snake. In this predator-prey pair, an evolutionary arms race has produced high levels of toxin in the newt and correspondingly high levels of toxin resistance in the snake.

Common garter snake (Thamnophis sirtalis sirtalis) has evolved
resistance to the defensive substance tetrodotoxin in its amphibian prey.

Cooperation

Not all co-evolved interactions between species involve conflict. Many cases of mutually beneficial interactions have evolved. For instance, an extreme cooperation exists between plants and the mycorrhizal fungi that grow on their roots and aid the plant in absorbing nutrients from the soil. This is a reciprocal relationship as the plants provide the fungi with sugars from photosynthesis. Here, the fungi actually grow inside plant cells, allowing them to exchange nutrients with their hosts, while sending signals that suppress the plant immune system.

Coalitions between organisms of the same species have also evolved. An extreme case is the eusociality found in social insects, such as bees, termites and ants, where sterile insects feed and guard the small number of organisms in a colony that are able to reproduce. On an even smaller scale, the somatic cells that make up the body of an animal limit their reproduction so they can maintain a stable organism, which then supports a small number of the animal's germ cells to produce offspring. Here, somatic cells respond to specific signals that instruct them whether to grow, remain as they are, or die. If cells ignore these signals and multiply inappropriately, their uncontrolled growth causes cancer.

Such cooperation within species may have evolved through the process of kin selection, which is where one organism acts to help raise a relative's offspring. This activity is selected for because if the helping individual contains alleles which promote the helping activity, it is likely that its kin will also contain these alleles and thus those alleles will be passed on. Other processes that may promote cooperation include group selection, where cooperation provides benefits to a group of organisms.

Speciation

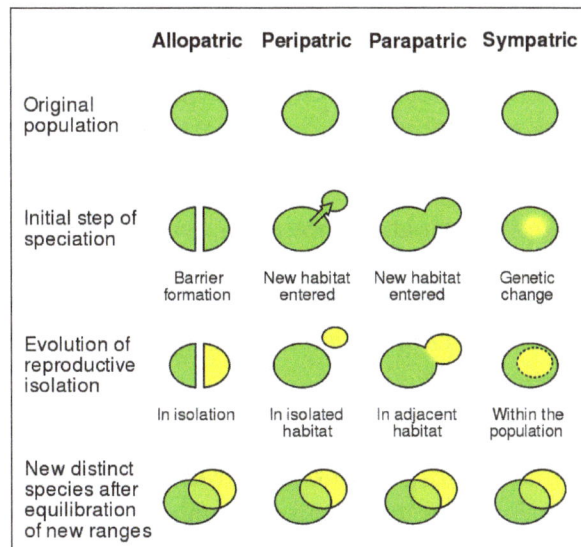

The four geographic modes of speciation.

Speciation is the process where a species diverges into two or more descendant species.

There are multiple ways to define the concept of "species." The choice of definition is dependent on the particularities of the species concerned. For example, some species concepts apply more readily toward sexually reproducing organisms while others lend themselves better toward asexual organisms. Despite the diversity of various species concepts, these various concepts can be placed into one of three broad philosophical approaches: interbreeding, ecological and phylogenetic. The Biological Species Concept (BSC) is a classic example of the interbreeding approach. Defined by evolutionary biologist Ernst Mayr in 1942, the BSC states that "species are groups of actually or potentially interbreeding natural populations, which are reproductively isolated from other such groups." Despite its wide and long-term use, the BSC like others is not without controversy, for example because these concepts cannot be applied to prokaryotes, and this is called the species problem. Some researchers have attempted a unifying monistic definition of species, while others adopt a pluralistic approach and suggest that there may be different ways to logically interpret the definition of a species.

Barriers to reproduction between two diverging sexual populations are required for the populations to become new species. Gene flow may slow this process by spreading the new genetic variants also to the other populations. Depending on how far two species have diverged since their most recent common ancestor, it may still be possible for them to produce offspring, as with horses and donkeys mating to produce mules. Such hybrids are generally infertile. In this case, closely related species may regularly interbreed, but hybrids will be selected against and the species will remain distinct. However, viable hybrids are occasionally formed and these new species can either have properties intermediate between their parent species, or possess a totally new phenotype. The importance of hybridisation in producing new species of animals is unclear, although cases have been seen in many types of animals, with the gray tree frog being a particularly well-studied example.

Speciation has been observed multiple times under both controlled laboratory conditions and in nature. In sexually reproducing organisms, speciation results from reproductive isolation followed by genealogical divergence. There are four primary geographic modes of speciation. The most common in animals is allopatric speciation, which occurs in populations initially isolated geographically, such as by habitat fragmentation or migration. Selection under these conditions can produce very rapid changes in the appearance and behaviour of organisms. As selection and drift act independently on populations isolated from the rest of their species, separation may eventually produce organisms that cannot interbreed.

The second mode of speciation is peripatric speciation, which occurs when small populations of organisms become isolated in a new environment. This differs from allopatric speciation in that the isolated populations are numerically much smaller than the parental population. Here, the founder effect causes rapid speciation after an increase in inbreeding increases selection on homozygotes, leading to rapid genetic change.

The third mode is parapatric speciation. This is similar to peripatric speciation in that a small population enters a new habitat, but differs in that there is no physical separation between these two populations. Instead, speciation results from the evolution of mechanisms that reduce gene flow between the two populations. Generally this occurs when there has been a drastic change in the environment within the parental species' habitat. One example is the grass Anthoxanthum odoratum, which can undergo parapatric speciation in response to localised metal pollution from mines. Here, plants evolve that have resistance to high levels of metals in the soil. Selection against interbreeding with the metal-sensitive parental population produced a gradual change in the flowering time of the metal-resistant plants, which eventually produced complete reproductive isolation. Selection against hybrids between the two populations may cause reinforcement, which is the evolution of traits that promote mating within a species, as well as character displacement, which is when two species become more distinct in appearance.

Finally, in sympatric speciation species diverge without geographic isolation or changes in habitat. This form is rare since even a small amount of gene flow may remove genetic differences between parts of a population. Generally, sympatric speciation in animals requires the evolution of both genetic differences and nonrandom mating, to allow reproductive isolation to evolve.

One type of sympatric speciation involves crossbreeding of two related species to produce a new hybrid species. This is not common in animals as animal hybrids are usually sterile. This is because during meiosis the homologous chromosomes from each parent are from different species and cannot successfully pair. However, it is more common in plants because plants often double their number of chromosomes, to form polyploids. This allows the chromosomes from each parental species to form matching pairs during meiosis, since each parent's chromosomes are represented by a pair already. An example of such a speciation event is when the plant species Arabidopsis thaliana and Arabidopsis arenosa crossbred to give the new species Arabidopsis suecica. This happened about 20,000 years ago, and the speciation process has been repeated in the laboratory, which allows the study of the genetic mechanisms involved in this process. Indeed, chromosome doubling within a species may be a common cause of reproductive isolation, as half the doubled chromosomes will be unmatched when breeding with undoubled organisms.

Speciation events are important in the theory of punctuated equilibrium, which accounts for the pattern in the fossil record of short "bursts" of evolution interspersed with relatively long periods of stasis, where species remain relatively unchanged. In this theory, speciation and rapid evolution are linked, with natural selection and genetic drift acting most strongly on organisms undergoing speciation in novel habitats or small populations. As a result, the periods of stasis in the fossil record correspond to the parental population and the organisms undergoing speciation and rapid evolution are found in small populations or geographically restricted habitats and therefore rarely being preserved as fossils.

Extinction

Tyrannosaurus rex. Non-avian dinosaurs died out in the Cretaceous–Paleogene extinctionevent at the end of the Cretaceous period.

Extinction is the disappearance of an entire species. Extinction is not an unusual event, as species regularly appear through speciation and disappear through extinction. Nearly all animal and plant species that have lived on Earth are now extinct, and extinction appears to be the ultimate fate of all species. These extinctions have happened continuously throughout the history of life, although the rate of extinction spikes in occasional mass extinction events. The Cretaceous–Paleogene extinction event, during which the non-avian dinosaurs became extinct, is the most well-known, but the earlier Permian–Triassic extinction event was even more severe, with approximately 96% of all marine species driven to extinction. The Holocene extinction event is an ongoing mass extinction associated with humanity's expansion across the globe over the past few thousand years. Present-day extinction rates are 100–1000 times greater than the background rate and up to 30% of current species may be extinct by the mid 21st century. Human activities are now the primary cause of the ongoing extinction event; global warming may further accelerate it in the future. Despite the estimated extinction of more than 99 percent of all species that ever lived on Earth, about 1 trillion species are estimated to be on Earth currently with only one-thousandth of one percent described.

The role of extinction in evolution is not very well understood and may depend on which type of extinction is considered. The causes of the continuous "low-level" extinction events, which form the majority of extinctions, may be the result of competition between species for limited resources (the competitive exclusion principle). If one species can out-compete another, this could produce species selection, with the fitter species surviving and the other species being driven to extinction. The intermittent mass extinctions are also important, but instead of acting as a selective force, they drastically reduce diversity in a nonspecific manner and promote bursts of rapid evolution and speciation in survivors.

Evolutionary History of Life

Origin of Life

The Earth is about 4.54 billion years old. The earliest undisputed evidence of life on Earth dates from at least 3.5 billion years ago, during the Eoarchean Era after a geological crust started to solidify following the earlier molten Hadean Eon. Microbial mat fossils have been found in 3.48 billion-year-old sandstone in Western Australia. Other early physical evidence of a biogenic substance is graphite in 3.7 billion-year-old metasedimentary rocks discovered in Western Greenland as well as "remains of biotic life" found in 4.1 billion-year-old rocks in Western Australia. Commenting on the Australian findings, Stephen Blair Hedges wrote, "If life arose relatively quickly on Earth, then it could be common in the universe." In July 2016, scientists reported identifying a set of 355 genes from the last universal common ancestor (LUCA) of all organisms living on Earth.

More than 99 percent of all species, amounting to over five billion species, that ever lived on Earth are estimated to be extinct. Estimates on the number of Earth's current species range from 10 million to 14 million, of which about 1.9 million are estimated to have been named and 1.6 million documented in a central database to date, leaving at least 80 percent not yet described.

Highly energetic chemistry is thought to have produced a self-replicating molecule around 4 billion years ago, and half a billion years later the last common ancestor of all life existed. The current scientific consensus is that the complex biochemistry that makes up life came from simpler chemical reactions. The beginning of life may have included self-replicating molecules such as RNA and the assembly of simple cells.

Common Descent

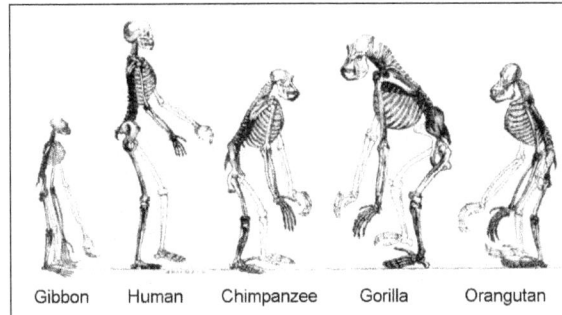

The hominoids are descendants of a common ancestor.

All organisms on Earth are descended from a common ancestor or ancestral gene pool. Current species are a stage in the process of evolution, with their diversity the product of a long series of speciation and extinction events. The common descent of organisms was first deduced from four simple facts about organisms: First, they have geographic distributions that cannot be explained by local adaptation. Second, the diversity of life is not a set of completely unique organisms, but organisms that share morphological similarities. Third, vestigial traits with no clear purpose resemble functional ancestral traits and finally, that organisms can be classified using these similarities into a hierarchy of nested groups—similar to a family tree. However, modern research has suggested that, due to horizontal gene transfer, this "tree of life" may be more complicated than a simple branching tree since some genes have spread independently between distantly related species.

Past species have also left records of their evolutionary history. Fossils, along with the comparative anatomy of present-day organisms, constitute the morphological, or anatomical, record. By comparing the anatomies of both modern and extinct species, paleontologists can infer the lineages of those species. However, this approach is most successful for organisms that had hard body parts, such as shells, bones or teeth. Further, as prokaryotes such as bacteria and archaea share a limited set of common morphologies, their fossils do not provide information on their ancestry.

More recently, evidence for common descent has come from the study of biochemical similarities between organisms. For example, all living cells use the same basic set of nucleotides and amino acids. The development of molecular genetics has revealed the record of evolution left in organisms' genomes: dating when species diverged through the molecular clock produced by mutations. For example, these DNA sequence comparisons have revealed that humans and chimpanzees share 98% of their genomes and analysing the few areas where they differ helps shed light on when the common ancestor of these species existed.

Evolution of Life

Prokaryotes inhabited the Earth from approximately 3–4 billion years ago. No obvious changes in morphology or cellular organisation occurred in these organisms over the next few billion years. The eukaryotic cells emerged between 1.6–2.7 billion years ago. The next major change in cell structure came when bacteria were engulfed by eukaryotic cells, in a cooperative association called endosymbiosis. The engulfed bacteria and the host cell then underwent coevolution, with the bacteria evolving into either mitochondria or hydrogenosomes. Another engulfment of cyanobacterial-like organisms led to the formation of chloroplasts in algae and plants.

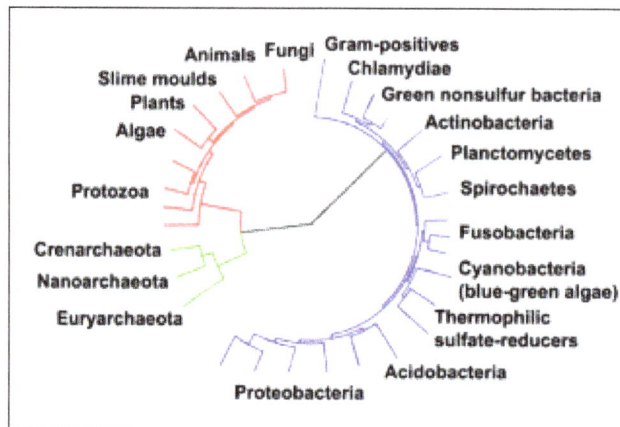

Evolutionary tree showing the divergence of modern species from their common ancestor in the centre. The three domains are coloured, with bacteria blue, archaea green and eukaryotes red.

The history of life was that of the unicellular eukaryotes, prokaryotes and archaea until about 610 million years ago when multicellular organisms began to appear in the oceans in the Ediacaran period. The evolution of multicellularity occurred in multiple independent events, in organisms as diverse as sponges, brown algae, cyanobacteria, slime moulds and myxobacteria. In January 2016, scientists reported that, about 800 million years ago, a minor genetic change in a single molecule called GK-PID may have allowed organisms to go from a single cell organism to one of many cells.

Soon after the emergence of these first multicellular organisms, a remarkable amount of biological diversity appeared over approximately 10 million years, in an event called the Cambrian explosion. Here, the majority of types of modern animals appeared in the fossil record, as well as unique lineages that subsequently became extinct. Various triggers for the Cambrian explosion have been proposed, including the accumulation of oxygen in the atmosphere from photosynthesis.

About 500 million years ago, plants and fungi colonised the land and were soon followed by arthropods and other animals. Insects were particularly successful and even today make up the majority of animal species. Amphibians first appeared around 364 million years ago, followed by early amniotes and birds around 155 million years ago (both from "reptile"-like lineages), mammals around 129 million years ago, homininae around 10 million years ago and modern humans around 250,000 years ago. However, despite the evolution of these large animals, smaller organisms similar to the types that evolved early in this process continue to be highly successful and dominate the Earth, with the majority of both biomass and species being prokaryotes.

Applications

Concepts and models used in evolutionary biology, such as natural selection, have many applications.

Artificial selection is the intentional selection of traits in a population of organisms. This has been used for thousands of years in the domestication of plants and animals. More recently, such selection has become a vital part of genetic engineering, with selectable markers such as antibiotic resistance genes being used to manipulate DNA. Proteins with valuable properties have evolved by repeated rounds of mutation and selection (for example modified enzymes and new antibodies) in a process called directed evolution.

Understanding the changes that have occurred during an organism's evolution can reveal the genes needed to construct parts of the body, genes which may be involved in human genetic disorders. For example, the Mexican tetra is an albino cavefish that lost its eyesight during evolution. Breeding together different populations of this blind fish produced some offspring with functional eyes, since different mutations had occurred in the isolated populations that had evolved in different caves. This helped identify genes required for vision and pigmentation.

Evolutionary theory has many applications in medicine. Many human diseases are not static phenomena, but capable of evolution. Viruses, bacteria, fungi and cancers evolve to be resistant to host immune defences, as well as pharmaceutical drugs. These same problems occur in agriculture with pesticide and herbicide resistance. It is possible that we are facing the end of the effective life of most of available antibiotics and predicting the evolution and evolvability of our pathogens and devising strategies to slow or circumvent it is requiring deeper knowledge of the complex forces driving evolution at the molecular level.

In computer science, simulations of evolution using evolutionary algorithms and artificial life started in the 1960s and were extended with simulation of artificial selection. Artificial evolution became a widely recognised optimisation method as a result of the work of Ingo Rechenberg in the 1960s. He used evolution strategies to solve complex engineering problems. Genetic algorithms in particular became popular through the writing of John Henry Holland. Practical applications also include automatic evolution of computer programmes. Evolutionary algorithms are now used to solve multi-dimensional problems more efficiently than software produced by human designers and also to optimise the design of systems.

Social and Cultural Responses

In the 19th century, particularly after the publication of "On the Origin of Species" in 1859, the idea that life had evolved was an active source of academic debate centred on the philosophical, social and religious implications of evolution. Today, the modern evolutionary synthesis is accepted by a vast majority of scientists. However, evolution remains a contentious concept for some theists.

While various religions and denominations have reconciled their beliefs with evolution through concepts such as theistic evolution, there are creationists who believe that evolution is contradicted by the creation myths found in their religions and who raise various objections to evolution. As had been demonstrated by responses to the publication of Vestiges of the Natural History of Creation in 1844, the most controversial aspect of evolutionary biology is the implication of human evolution that humans share common ancestry with apes and that the mental and moral faculties of humanity have the same types of natural causes as other inherited traits in animals. In some countries, notably the United States, these tensions between science and religion have fuelled the current creation–evolution controversy, a religious conflict focusing on politics and public education. While other scientific fields such as cosmology and Earth science also conflict with literal interpretations of many religious texts, evolutionary biology experiences significantly more opposition from religious literalists.

The teaching of evolution in American secondary school biology classes was uncommon in most of the first half of the 20th century. The Scopes Trial decision of 1925 caused the subject to become very rare in American secondary biology textbooks for a generation, but it was gradually re-introduced later and became legally protected with the 1968 Epperson v. Arkansas decision. Since then, the competing religious belief of creationism was legally disallowed in secondary school curricula in various decisions in the 1970s and 1980s, but it returned in pseudoscientific form as intelligent design (ID), to be excluded once again in the 2005 Kitzmiller v. Dover Area School District case. The debate over Darwin's ideas did not generate significant controversy in China.

ADAPTATION

In biology, adaptation has three related meanings. Firstly, it is the dynamic evolutionary process that fits organisms to their environment, enhancing their evolutionary fitness. Secondly, it is a state reached by the population during that process. Thirdly, it is a phenotypic trait or adaptive trait, with a functional role in each individual organism, that is maintained and has evolved through natural selection.

Organisms face a succession of environmental challenges as they grow, and show adaptive plasticity as traits develop in response to the imposed conditions. This gives them resilience to varying environments.

What Adaptation is?

Adaptation is primarily a process rather than a physical form or part of a body. An internal parasite (such as a liver fluke) can illustrate the distinction: such a parasite may have a very simple bodily structure, but nevertheless the organism is highly adapted to its specific environment. From this we see that adaptation is not just a matter of visible traits: in such parasites critical adaptations take place in the life cycle, which is often quite complex. However, as a practical term, "adaptation" often refers to a product: those features of a species which result from the process. Many aspects of an animal or plant can be correctly called adaptations, though there are always some features whose function remains in doubt. By using the term adaptation for the evolutionary process, and adaptive trait for the bodily part or function (the product), one may distinguish the two different senses of the word.

Adaptation is one of the two main processes that explain the observed diversity of species, such as the different species of Darwin's finches. The other process is speciation, in which new species arise, typically through reproductive isolation. A favourite example used today to study the interplay of adaptation and speciation is the evolution of cichlid fish in African lakes, where the question of reproductive isolation is complex.

Adaptation is not always a simple matter where the ideal phenotype evolves for a given external environment. An organism must be viable at all stages of its development and at all stages of its evolution. This places constraints on the evolution of development, behaviour, and structure of organisms. The main constraint, over which there has been much debate, is the requirement that each genetic and phenotypic change during evolution should be relatively small, because developmental systems are so complex and interlinked. However, it is not clear what "relatively small" should mean, for example polyploidy in plants is a reasonably common large genetic change. The origin of eukaryotic endosymbiosis is a more dramatic example.

All adaptations help organisms survive in their ecological niches. The adaptive traits may be structural, behavioural or physiological. Structural adaptations are physical features of an organism, such as shape, body covering, armament, and internal organization. Behavioural adaptations are inherited systems of behaviour, whether inherited in detail as instincts, or as a neuropsychological capacity for learning. Examples include searching for food, mating, and vocalizations. Physiological adaptations permit the organism to perform special functions such as making venom, secreting slime, and phototropism), but also involve more general functions such as growth and development, temperature regulation, ionic balance and other aspects of homeostasis. Adaptation affects all aspects of the life of an organism.

The following definitions are given by the evolutionary biologist Theodosius Dobzhansky:

- Adaptation is the evolutionary process whereby an organism becomes better able to live in its habitat or habitats.

- Adaptedness is the state of being adapted: the degree to which an organism is able to live and reproduce in a given set of habitats.

- An adaptive trait is an aspect of the developmental pattern of the organism which enables or enhances the probability of that organism surviving and reproducing.

What Adaptation is not?

Some generalists, such as birds, have the flexibility to adapt to urban areas.

Adaptation differs from flexibility, acclimatization, and learning. Flexibility deals with the relative capacity of an organism to maintain itself in different habitats: its degree of specialization. Acclimatization describes automatic physiological adjustments during life; learning means improvement in behavioral performance during life. These terms are preferred to adaptation for changes during life which are not inherited by the next generation.

Flexibility stems from phenotypic plasticity, the ability of an organism with a given genotype to change its phenotype in response to changes in its habitat, or to move to a different habitat. The degree of flexibility is inherited, and varies between individuals. A highly specialized animal or plant lives only in a well-defined habitat, eats a specific type of food, and cannot survive if its needs are not met. Many herbivores are like this; extreme examples are koalas which depend on Eucalyptus, and giant pandas which require bamboo. A generalist, on the other hand, eats a range of food, and can survive in many different conditions. Examples are humans, rats, crabs and many carnivores. The tendency to behave in a specialized or exploratory manner is inherited—it is an adaptation. Rather different is developmental flexibility: "An animal or plant is developmentally flexible if when it is raised in or transferred to new conditions, it changes in structure so that it is better fitted to survive in the new environment," writes evolutionary biologist John Maynard Smith.

If humans move to a higher altitude, respiration and physical exertion become a problem, but after spending time in high altitude conditions they acclimatize to the reduced partial pressure of oxygen, such as by producing more red blood cells. The ability to acclimatize is an adaptation, but the acclimatization itself is not. Fecundity goes down, but deaths from some tropical diseases also go down. Over a longer period of time, some people are better able to reproduce at high altitudes than others. They contribute more heavily to later generations, and gradually by natural selection the whole population becomes adapted to the new conditions. This has demonstrably occurred, as the observed performance of long-term communities at higher altitude is significantly better than the performance of new arrivals, even when the new arrivals have had time to acclimatize.

Adaptedness and Fitness

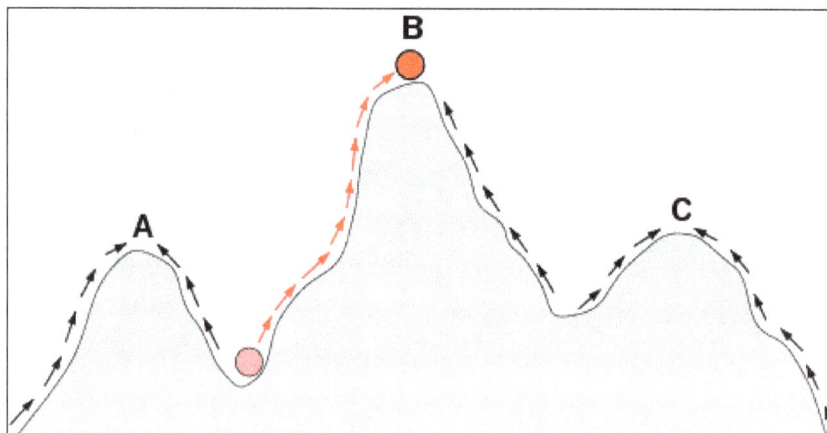

In this sketch of a fitness landscape, a population can evolve by following the arrows to the adaptive peak at point B, and the points A and C are local optima where a population could become trapped.

There is a relationship between adaptedness and the concept of fitness used in population genetics. Differences in fitness between genotypes predict the rate of evolution by natural selection. Natural selection changes the relative frequencies of alternative phenotypes, insofar as they are heritable. However, a phenotype with high adaptedness may not have high fitness. Dobzhansky mentioned the example of the Californian redwood, which is highly adapted, but a relict species in danger of extinction. Elliott Sober commented that adaptation was a retrospective concept since it implied something about the history of a trait, whereas fitness predicts a trait's future.

- Relative fitness: The average contribution to the next generation by a genotype or a class of genotypes, relative to the contributions of other genotypes in the population. This is also known as Darwinian fitness, selection coefficient, and other terms.

- Absolute fitness: The absolute contribution to the next generation by a genotype or a class of genotypes. Also known as the Malthusian parameter when applied to the population as a whole.

- Adaptedness: The extent to which a phenotype fits its local ecological niche. Researchers can sometimes test this through a reciprocal transplant.

Sewall Wright proposed that populations occupy adaptive peaks on a fitness landscape. To evolve to another, higher peak, a population would first have to pass through a valley of maladaptive intermediate stages, and might be "trapped" on a peak that is not optimally adapted.

Genetic Basis

A large diversity of genome DNAs in a species is the basis for adaptation and differentiation. A large population is needed to carry sufficient diversity. According to the misrepair-accumulation aging theory, the misrepair mechanism is important in maintaining a sufficient number of individuals in a species. Misrepair is a way of repair for increasing the surviving chance of an organism when it has severe injuries. Without misrepairs, no individual could survive to reproduction age. Thus misrepair mechanism is an essential mechanism for the survival of a species and for maintaining the number of individuals. Although individuals die from aging, genome DNAs are being recopied and transmitted by individuals generation by generation. In addition, the DNA misrepairs in germ cells contribute also to the diversity of genome DNAs.

Types

Changes in Habitat

Before Darwin, adaptation was seen as a fixed relationship between an organism and its habitat. It was not appreciated that as the climate changed, so did the habitat; and as the habitat changed, so did the biota. Also, habitats are subject to changes in their biota: for example, invasions of species from other areas. The relative numbers of species in a given habitat are always changing. Change is the rule, though much depends on the speed and degree of the change. When the habitat changes, three main things may happen to a resident population: habitat tracking, genetic change or extinction. In fact, all three things may occur in sequence. Of these three effects only genetic change brings about adaptation. When a habitat changes, the resident population typically moves to more suitable places; this is the typical response of flying insects or oceanic organisms, which have wide (though not unlimited) opportunity for movement. This common response is called habitat tracking. It is one explanation put forward for the periods of apparent stasis in the fossil record (the punctuated equilibrium theory).

Genetic Change

Genetic change occurs in a population when natural selection and mutations act on its genetic variability. The first pathways of enzyme-based metabolism may have been parts of purine nucleotide metabolism, with previous metabolic pathways being part of the ancient RNA world. By this means, the population adapts genetically to its circumstances. Genetic changes may result in visible structures, or may adjust physiological activity in a way that suits the habitat.

Habitats and biota do frequently change. Therefore, it follows that the process of adaptation is never finally complete. Over time, it may happen that the environment changes little, and the species comes to fit its surroundings better and better. On the other hand, it may happen that changes in the environment occur relatively rapidly, and then the species becomes less and less well adapted. Seen like this, adaptation is a genetic tracking process, which goes on all the time to some extent, but especially when the population cannot or does not move to another, less hostile area. Given enough genetic change, as well as specific demographic conditions, an adaptation may be enough to bring a population back from the brink of extinction in a process called evolutionary rescue. Adaptation does affect, to some extent, every species in a particular ecosystem.

Leigh Van Valen thought that even in a stable environment, competing species constantly had to adapt to maintain their relative standing. This became known as the Red Queen hypothesis, as seen in host-parasite interaction.

Co-Adaptation

Pollinating insects are co-adapted with flowering plants.

In coevolution, where the existence of one species is tightly bound up with the life of another species, new or 'improved' adaptations which occur in one species are often followed by the appearance and spread of corresponding features in the other species. These co-adaptational relationships are intrinsically dynamic, and may continue on a trajectory for millions of years, as has occurred in the relationship between flowering plants and pollinating insects.

Mimicry

Bates' work on Amazonian butterflies led him to develop the first scientific account of mimicry, especially the kind of mimicry which bears his name: Batesian mimicry. This is the mimicry by a palatable species of an unpalatable or noxious species, gaining a selective advantage. A common example seen in temperate gardens is the hoverfly, many of which—though bearing no sting—mimic the warning coloration of hymenoptera (wasps and bees). Such mimicry does not need to be perfect to improve the survival of the palatable species.

Bates, Wallace and Fritz Müller believed that Batesian and Müllerian mimicry provided evidence for the action of natural selection, a view which is now standard amongst biologists.

Trade-Offs

> "It is a profound truth that Nature does not know best; that genetical evolution... is a story of waste, makeshift, compromise and blunder."

— Peter Medawar

All adaptations have a downside: horse legs are great for running on grass, but they can't scratch their backs; mammals' hair helps temperature, but offers a niche for ectoparasites; the only flying penguins do is under water. Adaptations serving different functions may be mutually destructive. Compromise and makeshift occur widely, not perfection. Selection pressures pull in different directions, and the adaptation that results is some kind of compromise.

> "Since the phenotype as a whole is the target of selection, it is impossible to improve simultaneously all aspects of the phenotype to the same degree."

— Ernst Mayr

Consider the antlers of the Irish elk, (often supposed to be far too large; in deer antler size has an allometric relationship to body size). Obviously, antlers serve positively for defence against predators, and to score victories in the annual rut. But they are costly in terms of resource. Their size during the last glacial period presumably depended on the relative gain and loss of reproductive capacity in the population of elks during that time. As another example, camouflage to avoid detection is destroyed when vivid coloration is displayed at mating time. Here the risk to life is counterbalanced by the necessity for reproduction.

Stream-dwelling salamanders, such as Caucasian salamander or Gold-striped salamander have very slender, long bodies, perfectly adapted to life at the banks of fast small rivers and mountain brooks. Elongated body protects their larvae from being washed out by current. However, elongated body increases risk of desiccation and decreases dispersal ability of the salamanders; it also negatively affects their fecundity. As a result, fire salamander, less perfectly adapted to the mountain brook habitats, is in general more successful, have a higher fecundity and broader geographic range.

The peacock's ornamental train (grown anew in time for each mating season) is a famous adaptation. It must reduce his maneuverability and flight, and is hugely conspicuous; also, its growth costs food resources. Darwin's explanation of its advantage was in terms of sexual selection: "This depends on the advantage which certain individuals have over other individuals of the same sex and species, in exclusive relation to reproduction." The kind of sexual selection represented by the peacock is called 'mate choice,' with an implication that the process selects the more fit over the less fit, and so has survival value. The recognition of sexual selection was for a long time in abeyance, but has been rehabilitated.

The conflict between the size of the human foetal brain at birth, (which cannot be larger than about 400 cm3, else it will not get through the mother's pelvis) and the size needed for an adult brain (about 1400 cm3), means the brain of a newborn child is quite immature. The most vital things in human life (locomotion, speech) just have to wait while the brain grows and matures. That is the result of the birth compromise. Much of the problem comes from our upright bipedal stance, without which our pelvis could be shaped more suitably for birth. Neanderthals had a similar problem.

As another example, the long neck of a giraffe is a burden and a blessing. The neck of a giraffe can be up to 2 m (6 ft 7 in) in length. This neck can be used for inter-species competition or for foraging on tall trees where shorter herbivores cannot reach. However, as previously stated, there is always a trade-off. This long neck is heavy and it adds to the body mass of a giraffe, so the giraffe needs an abundance of nutrition to provide for this costly adaptation.

Shifts in Function

Pre-Adaptation

Pre-adaptation occurs when a population has characteristics which by chance are suited for a set of conditions not previously experienced. For example, the polyploid cordgrass Spartina townsendii is better adapted than either of its parent species to their own habitat of saline marsh and mud-flats. Among domestic animals, the White Leghorn chicken is markedly more resistant to vitamin B1 deficiency than other breeds; on a plentiful diet this makes no difference, but on a restricted diet this preadaptation could be decisive.

Pre-adaptation may arise because a natural population carries a huge quantity of genetic variability. In diploid eukaryotes, this is a consequence of the system of sexual reproduction, where mutant alleles get partially shielded, for example, by genetic dominance. Microorganisms, with their huge populations, also carry a great deal of genetic variability. The first experimental evidence of the pre-adaptive nature of genetic variants in microorganisms was provided by Salvador Luria and Max Delbrück who developed the Fluctuation Test, a method to show the random fluctuation of pre-existing genetic changes that conferred resistance to bacteriophages in Escherichia coli.

Co-Option of Existing Traits: Exaptation

The feathers of Sinosauropteryx, a dinosaur with feathers, were used for insulation, making them an exaptation for flight.

Features that now appear as adaptations sometimes arose by co-option of existing traits, evolved for some other purpose. The classic example is the ear ossicles of mammals, which we know from paleontological and embryological evidence originated in the upper and lower jaws and the hyoid bone of their synapsid ancestors, and further back still were part of the gill arches of early fish. The word exaptation was coined to cover these common evolutionary shifts in function. The flight feathers of birds evolved from the much earlier feathers of dinosaurs, which might have been used for insulation or for display.

Non-Adaptive Traits

Some traits do not appear to be adaptive, that is, they have a neutral or deleterious effect on fitness in the current environment. Because genes have pleiotropic effects, not all traits may be functional: they may be what Stephen Jay Gould and Richard Lewontin called spandrels, features brought about by neighbouring adaptations, like the triangular areas under neighbouring arches in architecture which began as functionless features.

Another possibility is that a trait may have been adaptive at some point in an organism's evolutionary history, but a change in habitats caused what used to be an adaptation to become unnecessary or even maladapted. Such adaptations are termed vestigial. Many organisms have vestigial organs, which are the remnants of fully functional structures in their ancestors. As a result of changes in lifestyle the organs became redundant, and are either not functional or reduced in functionality. Since any structure represents some kind of cost to the general economy of the body, an advantage may accrue from their elimination once they are not functional. Examples: wisdom teeth in humans; the loss of pigment and functional eyes in cave fauna; the loss of structure in endoparasites.

Extinction and Coextinction

If a population cannot move or change sufficiently to preserve its long-term viability, then obviously, it will become extinct, at least in that locale. The species may or may not survive in other locales. Species extinction occurs when the death rate over the entire species exceeds the birth rate for a long enough period for the species to disappear. It was an observation of Van Valen that groups of species tend to have a characteristic and fairly regular rate of extinction.

Just as there is co-adaptation, there is also coextinction, the loss of a species due to the extinction of another with which it is coadapted, as with the extinction of a parasitic insect following the loss of its host, or when a flowering plant loses its pollinator, or when a food chain is disrupted.

Adaptation raises philosophical issues concerning how biologists speak of function and purpose, as this carries implications of evolutionary history – that a feature evolved by natural selection for a specific reason – and potentially of supernatural intervention – that features and organisms exist because of a deity's conscious intentions. In his biology, Aristotle introduced teleology to describe the adaptedness of organisms, but without accepting the supernatural intention built into Plato's thinking, which Aristotle rejected. Modern biologists continue to face the same difficulty. On the one hand, adaptation is obviously purposeful: natural selection chooses what works and eliminates what does not. On the other hand, biologists want to deny conscious purpose in evolution. The dilemma gave rise to a famous joke by the evolutionary biologist Haldane: "Teleology is like a mistress to a biologist: he cannot live without her but he's unwilling to be seen with her in public.'" David Hull commented that Haldane's mistress "has become a lawfully wedded wife. Biologists no longer feel obligated to apologize for their use of teleological language; they flaunt it."

CONVERGENT EVOLUTION

Convergent evolution is the independent evolution of similar features in species of different lineages. Convergent evolution creates analogous structures that have similar form or function but were not present in the last common ancestor of those groups. The cladistic term for the same phenomenon is homoplasy. The recurrent evolution of flight is a classic example, as flying insects, birds, pterosaurs, and bats have independently evolved the useful capacity of flight. Functionally similar features that have arisen through convergent evolution are analogous, whereas homologous structures or traits have a common origin but can have dissimilar functions. Bird, bat, and pterosaur wings are analogous structures, but their forelimbs are homologous, sharing an ancestral state despite serving different functions.

The opposite of convergence is divergent evolution, where related species evolve different traits. Convergent evolution is similar to parallel evolution, which occurs when two independent species evolve in the same direction and thus independently acquire similar characteristics; for instance, gliding frogs have evolved in parallel from multiple types of tree frog.

Many instances of convergent evolution are known in plants, including the repeated development of C4 photosynthesis, seed dispersal by fleshy fruits adapted to be eaten by animals, and carnivory.

Two succulent plant genera, Euphorbia and Astrophytum, are only distantly related, but thespecies within each have converged on a similar body form.

Distinctions

Cladistics

In cladistics, a homoplasy is a trait shared by two or more taxa for any reason other than that they share a common ancestry. Taxa which do share ancestry are part of the same clade; cladistics seeks to arrange them according to their degree of relatedness to describe their phylogeny. Homoplastic traits caused by convergence are therefore, from the point of view of cladistics, confounding factors which could lead to an incorrect analysis.

Atavism

In some cases, it is difficult to tell whether a trait has been lost and then re-evolved convergently, or whether a gene has simply been switched off and then re-enabled later. Such a re-emerged trait is called an atavism. From a mathematical standpoint, an unused gene (selectively neutral) has a steadily decreasing probability of retaining potential functionality over time. The time scale of this process varies greatly in different phylogenies; in mammals and birds, there is a reasonable probability of remaining in the genome in a potentially functional state for around 6 million years.

Parallel vs. Convergent Evolution

When two species are similar in a particular character, evolution is defined as parallel if the ancestors were also similar, and convergent if they were not. Some scientists have argued that there is a continuum between parallel and convergent evolution, while others maintain that despite some overlap, there are still important distinctions between the two.

When the ancestral forms are unspecified or unknown, or the range of traits considered is not clearly specified, the distinction between parallel and convergent evolution becomes more subjective. For instance, the striking example of similar placental and marsupial forms is described by Richard Dawkins in The Blind Watchmaker as a case of convergent evolution, because mammals on each continent had a long evolutionary history prior to the extinction of the dinosaurs under which to accumulate relevant differences.

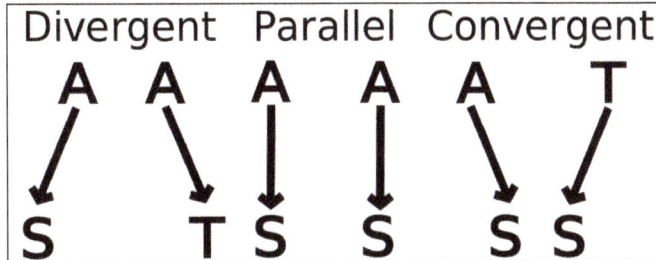

Evolution at an amino acid position. In each case, the left-hand species changes from having alanine (A) at a specific position in a protein in a hypothetical ancestor, and now has serine (S) there. The right-hand species may undergo divergent, parallel, or convergent evolution at this amino acid position relative to the first species.

At Molecular Level

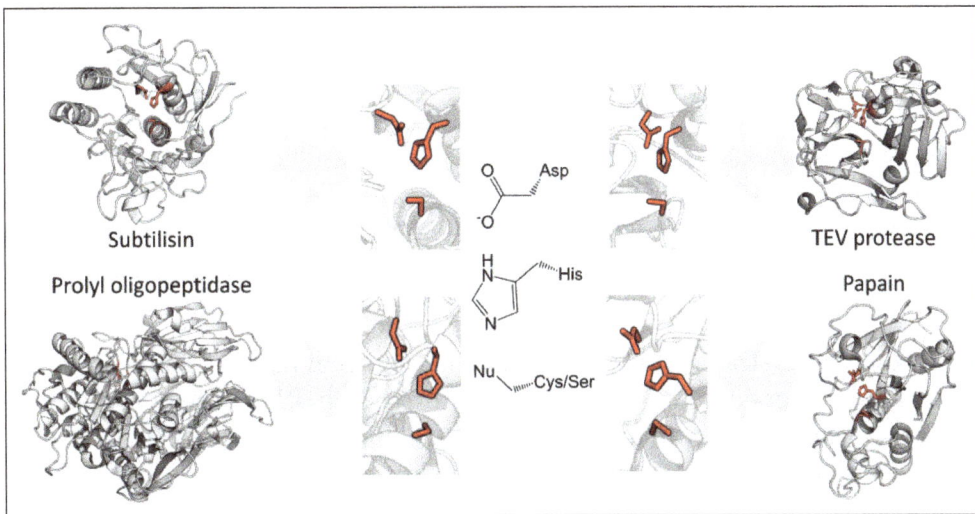

Evolutionary convergence of serine and cysteine protease towards the same catalytic triads organisation ofacid-base-nucleophile in different protease superfamilies. Shown are the triads of subtilisin, prolyl oligopeptidase, TEV protease, and papain.

Protease Active Sites

The enzymology of proteases provides some of the clearest examples of convergent evolution. These examples reflect the intrinsic chemical constraints on enzymes, leading evolution to converge on equivalent solutions independently and repeatedly.

Serine and cysteine proteases use different amino acid functional groups (alcohol or thiol) as a nucleophile. In order to activate that nucleophile, they orient an acidic and a basic residue in a catalytic triad. The chemical and physical constraints on enzyme catalysis have caused identical triad arrangements to evolve independently more than 20 times in different enzyme superfamilies.

Threonine proteases use the amino acid threonine as their catalytic nucleophile. Unlike cysteine and serine, threonine is a secondary alcohol (i.e. has a methyl group). The methyl group of threonine greatly restricts the possible orientations of triad and substrate, as the methyl clashes with either the enzyme backbone or the histidine base. Consequently, most threonine proteases use an N-terminal threonine in order to avoid such steric clashes. Several evolutionarily independent enzyme superfamilies with different protein folds use the N-terminal residue as a nucleophile. This commonality of active site but difference of protein fold indicates that the active site evolved convergently in those families.

Nucleic Acids

Convergence occurs at the level of DNA and the amino acid sequences produced by translating structural genes into proteins. Studies have found convergence in amino acid sequences in echolocating bats and the dolphin; among marine mammals; between giant and red pandas; and between the thylacine and canids. Convergence has also been detected in a type of non-coding DNA, cis-regulatory elements, such as in their rates of evolution; this could indicate either positive selection or relaxed purifying selection.

In Animal Morphology

Bodyplans

Swimming animals including fish such as herrings, marine mammals such as dolphins, and ichthyosaurs (of the Mesozoic) all converged on the same streamlined shape. The fusiform body-shape (a tube tapered at both ends) adopted by many aquatic animals is an adaptation to enable them to travel at high speed in a high drag environment. Similar body shapes are found in the earless seals and the eared seals: they still have four legs, but these are strongly modified for swimming.

The marsupial fauna of Australia and the placental mammals of the Old World have several strikingly similar forms, developed in two clades, isolated from each other. The body and especially the skull shape of the thylacine (Tasmanian tiger or Tasmanian wolf) converged with those of Canidae such as the red fox, Vulpes vulpes.

Dolphins and ichthyosaurs.

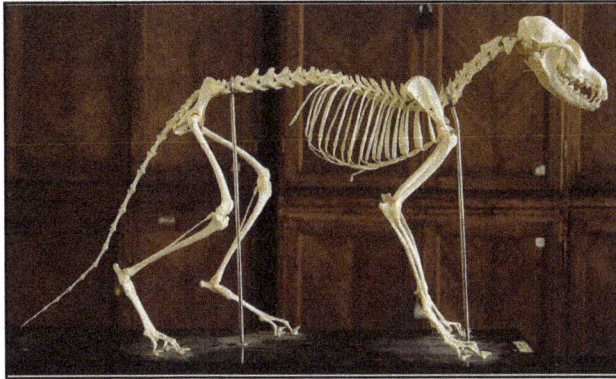

Red fox skeleton

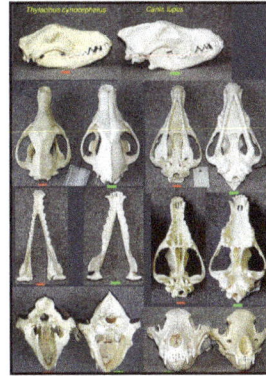

Skulls of thylacine (left), timber wolf (right)

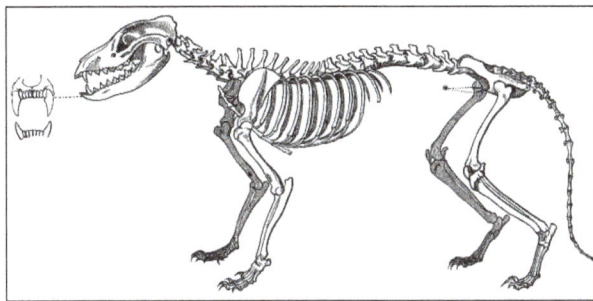

Thylacine skeleton

Echolocation

As a sensory adaptation, echolocation has evolved separately in cetaceans (dolphins and whales) and bats, but from the same genetic mutations.

Eyes

One of the best-known examples of convergent evolution is the camera eye of cephalopods (such as squid and octopus), vertebrates (including mammals) and cnidaria (such as jellyfish). Their last common ancestor had at most a simple photoreceptive spot, but a range of processes led to the progressive refinement of camera eyes — with one sharp difference: the cephalopod eye is "wired" in the opposite direction, with blood and nerve vessels entering from the back of the retina, rather than the front as in vertebrates. As a result, cephalopods lack a blind spot.

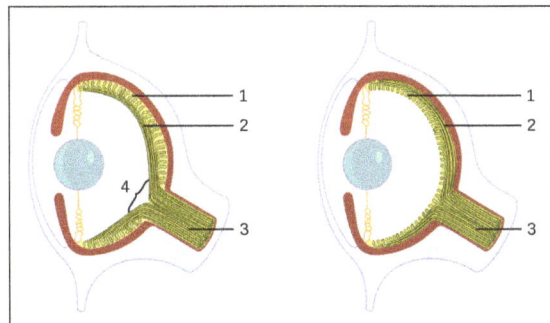

The camera eyes of vertebrates (left) and cephalopods (right) developed independently and are wired differently; for instance, optic nerve fibres reach the vertebrate retina from the front, creating a blind spot.

Flight

Birds and bats have homologous limbs because they are both ultimately derived from terrestrial tetrapods, but their flight mechanisms are only analogous, so their wings are examples of functional convergence. The two groups have powered flight, evolved independently. Their wings differ substantially in construction. The bat wing is a membrane stretched across four extremely elongated fingers and the legs. The airfoil of the bird wing is made of feathers, strongly attached to the forearm (the ulna) and the highly fused bones of the wrist and hand (the carpometacarpus), with only tiny remnants of two fingers remaining, each anchoring a single feather. So, while the wings of bats and birds are functionally convergent, they are not anatomically convergent. Birds and bats also share a high concentration of cerebrosides in the skin of their wings. This improves skin flexibility, a trait useful for flying animals; other mammals have a far lower concentration. The extinct pterosaurs independently evolved wings from their fore- and hindlimbs, while insects have wings that evolved separately from different organs.

Flying squirrels and sugar gliders are much alike in their body plans, with gliding wings stretched between their limbs, but flying squirrels are placental mammals while sugar gliders are marsupials, widely separated within the mammal lineage.

Hummingbird hawk-moths and hummingbirds have evolved similar flight and feeding patterns.

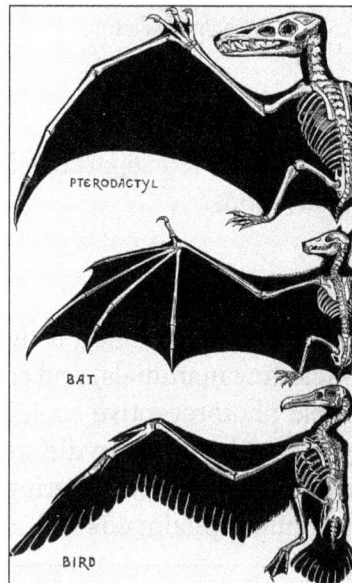

Vertebrate wings are partly homologous (from forelimbs), but analogous
as organs of flight in (1) pterosaurs, (2) bats, (3) birds, evolved separately.

Insect Mouthparts

Insect mouthparts show many examples of convergent evolution. The mouthparts of different insect groups consist of a set of homologous organs, specialised for the dietary intake of that insect group. Convergent evolution of many groups of insects led from original biting-chewing mouthparts to different, more specialised, derived function types. These include, for example, the proboscis of flower-visiting insects such as bees and flower beetles, or the biting-sucking mouthparts of blood-sucking insects such as fleas and mosquitos.

Opposable Thumbs

Opposable thumbs allowing the grasping of objects are most often associated with primates, like humans, monkeys, apes, and lemurs. Opposable thumbs also evolved in giant pandas, but these are completely different in structure, having six fingers including the thumb, which develops from a wrist bone entirely separately from other fingers.

Primates

Convergent evolution in humans includes blue eye colour and light skin colour. When humans migrated out of Africa, they moved to more northern latitudes with less intense sunlight. It was beneficial to them to reduce their skin pigmentation. It appears certain that there was some lightening of skin colour before European and East Asian lineages diverged, as there are some skin-lightening genetic differences that are common to both groups. However, after the lineages diverged and became genetically isolated, the skin of both groups lightened more, and that additional lightening was due to different genetic changes.

Lemurs and humans are both primates. Ancestral primates had brown eyes, as most primates do today. The genetic basis of blue eyes in humans has been studied in detail and much is known about it. It is not the case that one gene locus is responsible, say with brown dominant to blue eye colour. However, a single locus is responsible for about 80% of the variation. In lemurs, the differences between blue and brown eyes are not completely known, but the same gene locus is not involved.

Humans

Lemurs

Despite the similarity of appearance, the genetic basis of blue eyes is different in humans and lemurs.

Despite the similar lightening of skin colour after moving out of Africa, different genes wereinvolved in European (left) and East-Asian (right) lineages.

In Plants

Carbon Fixation

While convergent evolution is often illustrated with animal examples, it has often occurred in plant evolution. For instance, C4 photosynthesis, one of the three major carbon-fixing biochemical processes, has arisen independently up to 40 times. About 7,600 plant species of angiosperms use C4 carbon fixation, with many monocots including 46% of grasses such as maize and sugar cane, and dicots including several species in the Chenopodiaceae and the Amaranthaceae.

In myrmecochory, seeds such as those of Chelidonium majus have a hard coating and an attachedoil body, an elaiosome, for dispersal by ants.

Fruits

A good example of convergence in plants is the evolution of edible fruits such as apples. These pomes incorporate (five) carpels and their accessory tissues forming the apple's core, surrounded by structures from outside the botanical fruit, the receptacle or hypanthium. Other edible fruits include other plant tissues; for example, the fleshy part of a tomato is the walls of the pericarp. This implies convergent evolution under selective pressure, in this case the competition for seed dispersal by animals through consumption of fleshy fruits.

Seed dispersal by ants (myrmecochory) has evolved independently more than 100 times, and is present in more than 11,000 plant species. It is one of the most dramatic examples of convergent evolution in biology.

Carnivory

Carnivory has evolved multiple times independently in plants in widely separated groups. In three species studied, Cephalotus follicularis, Nepenthes alata and Sarracenia purpurea, there has been convergence at the molecular level. Carnivorous plants secrete enzymes into the digestive fluid they produce. By studying phosphatase, glycoside hydrolase, glucanase, RNAse and chitinase enzymes as well as a pathogenesis-related protein and a thaumatin-related protein, the authors found many convergent amino acid substitutions. These changes were not at the enzymes' catalytic sites, but rather on the exposed surfaces of the proteins, where they might interact with other components of the cell or the digestive fluid. The authors also found that homologous genes in the non-carnivorous plant Arabidopsis thaliana tend to have their expression increased when the plant is stressed, leading the authors to suggest that stress-responsive proteins have often been co-opted in the repeated evolution of carnivory.

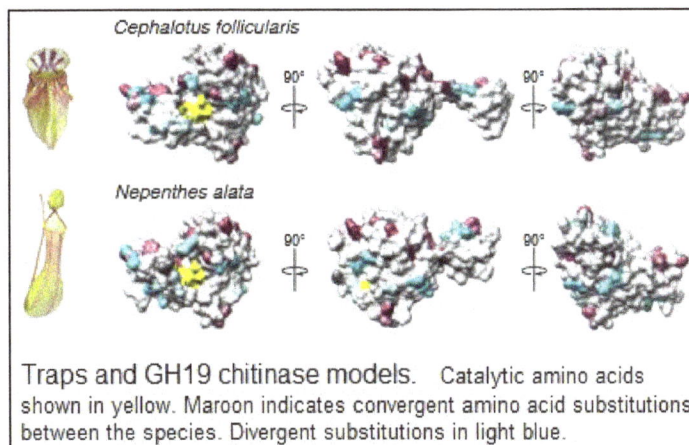

Traps and GH19 chitinase models. Catalytic amino acids shown in yellow. Maroon indicates convergent amino acid substitutions between the species. Divergent substitutions in light blue.

Molecular convergence in carnivorous plants.

Methods of Inference

Phylogenetic reconstruction and ancestral state reconstruction proceed by assuming that evolution has occurred without convergence. Convergent patterns may, however, appear at higher levels in a phylogenetic reconstruction, and are sometimes explicitly sought by investigators. The methods applied to infer convergent evolution depend on whether pattern-based or process-based convergence is expected. Pattern-based convergence is the broader term, for when two or more lineages independently evolve patterns of similar traits. Process-based convergence is when the convergence is due to similar forces of natural selection.

Pattern-Based Measures

Earlier methods for measuring convergence incorporate ratios of phenotypic and phylogenetic distance by simulating evolution with a Brownian motion model of trait evolution along a phylogeny. More recent methods also quantify the strength of convergence. One drawback to keep in mind is that these methods can confuse long-term stasis with convergence due to phenotypic similarities. Stasis occurs when there is little evolutionary change among taxa.

Distance-based measures assess the degree of similarity between lineages over time. Frequency-based measures assess the number of lineages that have evolved in a particular trait space.

Process-Based Measures

Methods to infer process-based convergence fit models of selection to a phylogeny and continuous trait data to determine whether the same selective forces have acted upon lineages. This uses the Ornstein-Uhlenbeck (OU) process to test different scenarios of selection. Other methods rely on an a priori specification of where shifts in selection have occurred.

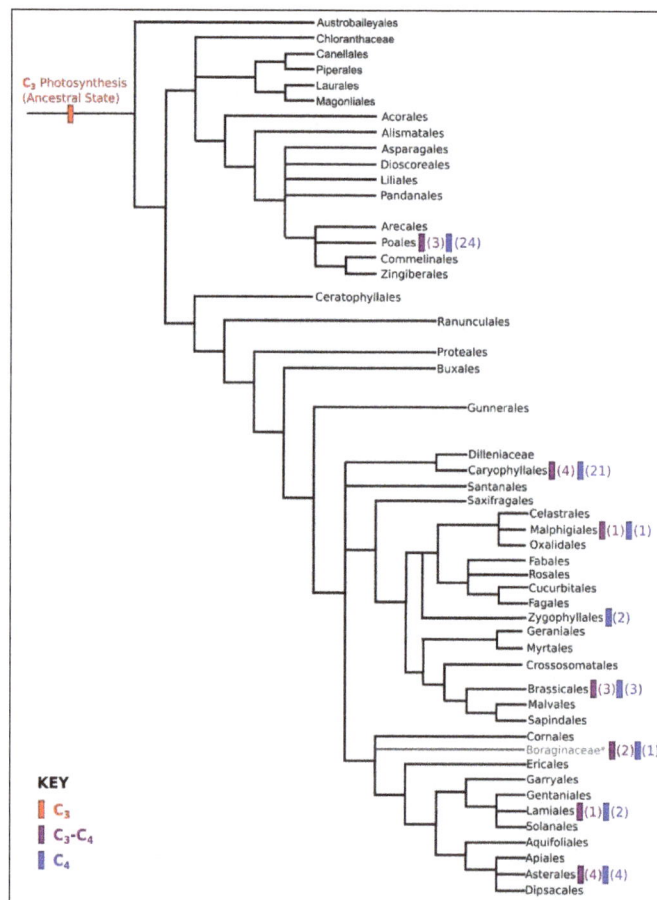

Angiosperm phylogeny of orders based on classification by the Angiosperm Phylogeny Group. The figure shows the number of inferred independent origins of C3-C4 photosynthesis and C4 photosynthesis in parentheses.

DIVERGENT EVOLUTION

Divergent evolution or divergent selection is the accumulation of differences between closely related populations within a species, leading to speciation. Divergent evolution is typically exhibited when two populations become separated by a geographic barrier (such as in allopatric or peripatric speciation) and experience different selective pressures that drive adaptations to their new environment. After many generations and continual evolution, the populations become unable to interbreed with one another. The American naturalist J. T. Gulick was the first to use the term "divergent evolution", with its use becoming widespread in modern evolutionary literature. Classic examples of divergence in nature are the adaptive radiation of the finches of the Galapagos or the coloration differences in populations of a species that live in different habitats such as with pocket mice and fence lizards.

The term can also be applied in molecular evolution, such as to proteins that derive from homologous genes. Both orthologous genes (resulting from a speciation event) and paralogous genes (resulting from gene duplication) can illustrate divergent evolution. Through gene duplication, it is possible for divergent evolution to occur between two genes within a species. Similarities between species that have diverged are due to their common origin, so such similarities are homologies. In contrast, convergent evolution arises when an adaptation has arisen independently, creating analogous structures such as the wings of birds and of insects.

Creation and Usage

The term divergent evolution is believed to have been first used by J. T. Gulick. Divergent evolution is commonly defined as what occurs when two groups of the same species evolve different traits within those groups in order to accommodate for differing environmental and social pressures. Various examples of such pressures can include predation, food supplies, and competition for mates. The tympanal ears of certain nocturnal insects are believed to be a result of needing the ultrasonic hearing that tympanal ears provide in order to hear predators in the dark. Non-nocturnal insects - that do not need to fear nocturnal predators - are often found to lack these tympanal ears.

Causes

Animals undergo divergent evolution for a number of reasons. Predators or their absence, changes in the environment, and the time at which certain animals are most active are chief among them.

Predators

A lack of predators – predatory birds and mammals - for cliff-side nest residing kittiwake caused that particular group of kittiwake to lose their ancestral mobbing behavior that had been exhibited up until that point for protecting young. The mobbing behavior normally displayed by the kittiwake is lost when the kittiwake take residence in this area with little threat from predators towards their young. The mobbing behavior was originally developed to protect ground-level nests containing young from various predators such as reptiles, mammals and other birds.

Environment

The cliff-side nesting area itself was similarly responsible for the kittiwakes losing their mobbing mentality – predatory mammals small enough to fit on the cliff edges along with the kittiwakes and their offspring would not be able to make the climb up while predatory birds would not be able to maneuver near the cliff face while also being afflicted by the weather conditions of the area.

Distinctions

Divergent evolution is always coupled with convergent evolution, as they are both similar and different in various facets such as whether something evolves, what evolves, and why it evolves. It is instructive to compare divergent evolution with both convergent and parallel evolution.

Divergent versus Convergent Evolution

Convergent evolution is defined as a similar trait evolution that occurs in two otherwise different species of animal as a result of those two species living in similar environments with similar environmental pressures like predators and food supply. It differs from divergent evolution in that the species involved are different while the traits they obtain do not differ from each other. An example of convergent evolution is the development of horns in various species for sparring over mates, resources, and territory.

Divergent versus Parallel Evolution

Parallel evolution is the development of a similar trait in species descending from the same ancestor. It is similar to divergent evolution in that the species descend from the same ancestor, but it differs in that the trait is the same while in divergent evolution the trait is not. An example of parallel evolution are certain arboreal frog species, 'flying' frogs, in both Old World families and New World families having developed the ability of gliding flight. They have "enlarged hands and feet, full webbing between all fingers and toes, lateral skin flaps on the arms and legs, and reduced weight per snout-vent length".

Darwin's Finches

One of the most famous examples of divergent evolution is the case of Darwin's Finches. During Darwin's travels to the Galápagos Islands he discovered several different species of finch that shared a common ancestor. They lived on varying diets and had beaks that differed in shape and size reflecting their diet. The change in beak shape and size was believed to be a result of the lengths the birds had to go to in order to support their change in diet. Some Galapagos finches have beaks that are larger and more powerful to crack nuts with. A different type allows the bird to use cactus spines to spear insects in the bark of trees.

Divergent Evolution in Dogs

Another good example of divergent evolution is the origin of the domestic dog and the modern wolf. Dogs and wolves both diverged from a common ancestor. To further support divergent evolution of dogs and wolves, genomic research was conducted to compare mitochondrial DNA to indicate the presence of shared ancestry. Taking 162 wolves from various parts of the world as well as 140 dogs of 60 different breeds, it is found that dogs and wolves have shared ancestry by how similar their DNA sequences are. Comparison of the physical characteristics reveal that dogs and wolves have similar body shape, skull size, and limb formation, further supporting their close genetic makeup and thus shared ancestry. An example of this would be how physically and behaviorally similar malamutes and huskies are to wolves. Huskies and malamutes have very similar body size and skull shape. Huskies and wolves share similar coat patterns as well as tolerance to cold. In the hypothetical situations, mutations and breeding events were simulated to show the progression of the wolf behavior over ten generations. The results concluded that even though the last generation of the wolves were more docile and less aggressive, the temperament of the wolves fluctuated greatly from one generation to the next.

MACROEVOLUTION

Macroevolution is evolution on a scale at or above the level of species, in contrast with micro-evolution, which refers to smaller evolutionary changes of allele frequencies within a species or population. Macroevolution and microevolution describe fundamentally identical processes on different scales.

The process of speciation may fall within the purview of either, depending on the forces thought to drive it. Paleontology, evolutionary developmental biology, comparative genomics and genomic phylostratigraphy contribute most of the evidence for macroevolution's patterns and processes.

Macroevolution and the Modern Synthesis

Within the modern synthesis of the early 20th century, macroevolution is thought of as the compounded effects of microevolution. Thus, the distinction between micro- and macroevolution is not a fundamental one – the only difference between them is of time and scale. As Ernst W. Mayr observes, "transspecific evolution is nothing but an extrapolation and magnification of the events that take place within populations and species...it is misleading to make a distinction between the causes of micro- and macroevolution". However, time is not a necessary distinguishing factor – macroevolution can happen without gradual compounding of small changes; whole-genome duplication can result in speciation occurring over a single generation – this is especially common in plants.

Changes in the genes regulating development have also been proposed as being important in producing speciation through large and relatively sudden changes in animals' morphology.

Types of Macroevolution

There are many ways to view macroevolution, for example, by observing changes in the genetics, morphology, taxonomy, ecology, and behavior of organisms, though these are interrelated. Sahney et al. stated the connection as "As taxonomic diversity has increased, there have been incentives for tetrapods to move into new modes of life, where initially resources may seem unlimited, there are few competitors and possible refuge from danger. And as ecological diversity increases, taxa diversify from their ancestors at a much greater rate among faunas with more superior, innovative or more flexible adaptations."

Molecular evolution occurs through small changes in the molecular or cellular level. Over a long period of time, this can cause big effects on the genetics of organisms. Taxonomic evolution occurs through small changes between populations and then species. Over a long period of time, this can cause big effects on the taxonomy of organisms, with the growth of whole new clades above the species level. Morphological evolution occurs through small changes in the morphology of an organism. Over a long period of time, this can cause big effects on the morphology of major clades. This can be clearly seen in the Cetacea, where throughout the group's early evolution, hindlimbs were still present. However over millions of years the hindlimbs regressed and became internal.

Abrupt transformations from one biologic system to another, for example the passing of life from water into land or the transition from invertebrates to vertebrates, are rare. Few major biological types have emerged during the evolutionary history of life. When lifeforms take such giant leaps, they meet little to no competition and are able to exploit many available niches, following an adaptive radiation. This can lead to convergent evolution as the empty niches are filled by whichever lifeform encounters them.

Areas

Subjects studied within macroevolution include:

- Adaptive radiations such as the Cambrian Explosion.

- Changes in biodiversity through time.

- Genome evolution, like horizontal gene transfer, genome fusions in endosymbioses, and adaptive changes in genome size.

- Mass extinctions.

- Estimating diversification rates, including rates of speciation and extinction.

- The debate between punctuated equilibrium and gradualism.

- The role of development in shaping evolution, particularly such topics as heterochrony and phenotypic plasticity.

MICROEVOLUTION

Microevolution is the change in allele frequencies that occurs over time within a population. This change is due to four different processes: mutation, selection (natural and artificial), gene flow and genetic drift. This change happens over a relatively short (in evolutionary terms) amount of time compared to the changes termed macroevolution which is where greater differences in the population occur.

Population genetics is the branch of biology that provides the mathematical structure for the study of the process of microevolution. Ecological genetics concerns itself with observing microevolution in the wild. Typically, observable instances of evolution are examples of microevolution; for example, bacterial strains that have antibiotic resistance.

Microevolution over time leads to speciation or the appearance of novel structure, sometimes classified as macroevolution. Macro and microevolution describe fundamentally identical processes on different scales.

Difference from Macroevolution

Macroevolution and microevolution describe fundamentally identical processes on different time scales. Microevolution refers to small evolutionary changes (typically described as changes in allele frequencies) within a species or population. while macroevolution is evolution on a scale of separated gene pools. Macroevolutionary studies focus on change that occurs at or above the level of species.

Four Processes

Mutation

Mutations are changes in the DNA sequence of a cell's genome and are caused by radiation, viruses, transposons and mutagenic chemicals, as well as errors that occur during meiosis or DNA replication. Errors are introduced particularly often in the process of DNA replication, in the polymerization of the second strand. These errors can also be induced by the organism itself, by cellular processes such as hypermutation. Mutations can affect the phenotype of an organism, especially if they occur within the protein coding sequence of a gene. Error rates are usually very low—1 error in every 10–100 million bases—due to the proofreading ability of DNA polymerases. (Without proofreading error rates are a thousandfold higher; because many viruses rely on DNA and RNA polymerases that lack proofreading ability, they experience higher mutation rates.) Processes that increase the rate of changes in DNA are called mutagenic: mutagenic chemicals promote errors in DNA replication, often by interfering with the structure of base-pairing, while UV radiation induces mutations by causing damage to the DNA structure. Chemical damage to DNA occurs naturally as well, and cells use DNA repair mechanisms to repair mismatches and breaks in DNA—nevertheless, the repair sometimes fails to return the DNA to its original sequence.

In organisms that use chromosomal crossover to exchange DNA and recombine genes, errors in alignment during meiosis can also cause mutations. Errors in crossover are especially likely when similar sequences cause partner chromosomes to adopt a mistaken alignment making some regions in genomes more prone to mutating in this way. These errors create large structural changes in DNA sequence—duplications, inversions or deletions of entire regions, or the accidental exchanging of whole parts between different chromosomes (called translocation).

Mutation can result in several different types of change in DNA sequences; these can either have no effect, alter the product of a gene, or prevent the gene from functioning. Studies in the fly Drosophila melanogaster suggest that if a mutation changes a protein produced by a gene, this will probably be harmful, with about 70 percent of these mutations having damaging effects, and the remainder being either neutral or weakly beneficial. Due to the damaging effects that mutations can have on cells, organisms have evolved mechanisms such as DNA repair to remove mutations. Therefore, the optimal mutation rate for a species is a trade-off between costs of a high mutation rate, such as deleterious mutations, and the metabolic costs of maintaining systems to reduce the mutation rate, such as DNA repair enzymes. Viruses that use RNA as their genetic material have rapid mutation rates, which can be an advantage since these viruses will evolve constantly and rapidly, and thus evade the defensive responses of e.g. the human immune system.

Mutations can involve large sections of DNA becoming duplicated, usually through genetic recombination. These duplications are a major source of raw material for evolving new genes, with tens to hundreds of genes duplicated in animal genomes every million years. Most genes belong to larger families of genes of shared ancestry. Novel genes are produced by several methods, commonly through the duplication and mutation of an ancestral gene, or by recombining parts of different genes to form new combinations with new functions.

Here, domains act as modules, each with a particular and independent function, that can be mixed together to produce genes encoding new proteins with novel properties. For example, the human eye uses four genes to make structures that sense light: three for color vision and one for night vision; all four arose from a single ancestral gene. Another advantage of duplicating a gene (or even an entire genome) is that this increases redundancy; this allows one gene in the pair to acquire a new function while the other copy performs the original function. Other types of mutation occasionally create new genes from previously noncoding DNA.

Selection

Selection is the process by which heritable traits that make it more likely for an organism to survive and successfully reproduce become more common in a population over successive generations.

It is sometimes valuable to distinguish between naturally occurring selection, natural selection, and selection that is a manifestation of choices made by humans, artificial selection. This distinction is rather diffuse. Natural selection is nevertheless the dominant part of selection.

The natural genetic variation within a population of organisms means that some individuals will survive more successfully than others in their current environment. Factors which affect reproductive success are also important, an issue which Charles Darwin developed in his ideas on sexual selection.

Natural selection acts on the phenotype, or the observable characteristics of an organism, but the genetic (heritable) basis of any phenotype which gives a reproductive advantage will become more common in a population. Over time, this process can result in adaptations that specialize organisms for particular ecological niches and may eventually result in the speciation (the emergence of new species).

Natural selection is one of the cornerstones of modern biology. The term was introduced by Darwin in his groundbreaking 1859 book On the Origin of Species, in which natural selection was described by analogy to artificial selection, a process by which animals and plants with traits considered desirable by human breeders are systematically favored for reproduction. The concept of natural selection was originally developed in the absence of a valid theory of heredity; at the time of Darwin's writing, nothing was known of modern genetics. The union of traditional Darwinian evolution with subsequent discoveries in classical and molecular genetics is termed the modern evolutionary synthesis. Natural selection remains the primary explanation for adaptive evolution.

Genetic Drift

Genetic drift is the change in the relative frequency in which a gene variant (allele) occurs in a population due to random sampling. That is, the alleles in the offspring in the population are a random sample of those in the parents. And chance has a role in determining whether a given individual survives and reproduces. A population's allele frequency is the fraction or percentage of its gene copies compared to the total number of gene alleles that share a particular form.

Genetic drift is an evolutionary process which leads to changes in allele frequencies over time. It may cause gene variants to disappear completely, and thereby reduce genetic variability. In contrast to natural selection, which makes gene variants more common or less common depending on their reproductive success, the changes due to genetic drift are not driven by environmental or adaptive pressures, and may be beneficial, neutral, or detrimental to reproductive success.

The effect of genetic drift is larger in small populations, and smaller in large populations. Vigorous debates wage among scientists over the relative importance of genetic drift compared with natural selection. Ronald Fisher held the view that genetic drift plays at the most a minor role in evolution, and this remained the dominant view for several decades. In 1968 Motoo Kimura rekindled the debate with his neutral theory of molecular evolution which claims that most of the changes in the genetic material are caused by genetic drift. The predictions of neutral theory, based on genetic drift, do not fit recent data on whole genomes well: these data suggest that the frequencies of neutral alleles change primarily due to selection at linked sites, rather than due to genetic drift by means of sampling error.

Gene Flow

Gene flow is the exchange of genes between populations, which are usually of the same species. Examples of gene flow within a species include the migration and then breeding of organisms, or the exchange of pollen. Gene transfer between species includes the formation of hybrid organisms and horizontal gene transfer.

Migration into or out of a population can change allele frequencies, as well as introducing genetic variation into a population. Immigration may add new genetic material to the established gene pool of a population. Conversely, emigration may remove genetic material. As barriers to reproduction between two diverging populations are required for the populations to become new species, gene flow may slow this process by spreading genetic differences between the populations. Gene flow is hindered by mountain ranges, oceans and deserts or even man-made structures such as the Great Wall of China, which has hindered the flow of plant genes.

Depending on how far two species have diverged since their most recent common ancestor, it may still be possible for them to produce offspring, as with horses and donkeys mating to produce mules. Such hybrids are generally infertile, due to the two different sets of chromosomes being unable to pair up during meiosis. In this case, closely related species may regularly interbreed, but hybrids will be selected against and the species will remain distinct. However, viable hybrids are occasionally formed and these new species can either have properties intermediate between their parent species, or possess a totally new phenotype. The importance of hybridization in creating new species of animals is unclear, although cases have been seen in many types of animals, with the gray tree frog being a particularly well-studied example.

Hybridization is, however, an important means of speciation in plants, since polyploidy (having more than two copies of each chromosome) is tolerated in plants more readily than in animals. Polyploidy is important in hybrids as it allows reproduction, with the two different sets of chromosomes each being able to pair with an identical partner during meiosis. Polyploid hybrids also have more genetic diversity, which allows them to avoid inbreeding depression in small populations.

Horizontal gene transfer is the transfer of genetic material from one organism to another organism that is not its offspring; this is most common among bacteria. In medicine, this contributes to the spread of antibiotic resistance, as when one bacteria acquires resistance genes it can rapidly transfer them to other species. Horizontal transfer of genes from bacteria to eukaryotes such as the yeast Saccharomyces cerevisiae and the adzuki bean beetle Callosobruchus chinensis may also have occurred. An example of larger-scale transfers are the eukaryotic bdelloid rotifers, which appear to have received a range of genes from bacteria, fungi, and plants. Viruses can also carry DNA between organisms, allowing transfer of genes even across biological domains. Large-scale gene transfer has also occurred between the ancestors of eukaryotic cells and prokaryotes, during the acquisition of chloroplasts and mitochondria.

Gene flow is the transfer of alleles from one population to another.

Migration into or out of a population may be responsible for a marked change in allele frequencies. Immigration may also result in the addition of new genetic variants to the established gene pool of a particular species or population.

There are a number of factors that affect the rate of gene flow between different populations. One of the most significant factors is mobility, as greater mobility of an individual tends to give it greater migratory potential. Animals tend to be more mobile than plants, although pollen and seeds may be carried great distances by animals or wind.

Maintained gene flow between two populations can also lead to a combination of the two gene pools, reducing the genetic variation between the two groups. It is for this reason that gene flow strongly acts against speciation, by recombining the gene pools of the groups, and thus, repairing the developing differences in genetic variation that would have led to full speciation and creation of daughter species.

For example, if a species of grass grows on both sides of a highway, pollen is likely to be transported from one side to the other and vice versa. If this pollen is able to fertilise the plant where it ends up and produce viable offspring, then the alleles in the pollen have effectively been able to move from the population on one side of the highway to the other.

References

- Van Bael, Sunshine A; et al. (April 2008). "Birds as predators in tropical agroforestry systems". Ecology. 89 (4): 928–934. Doi:10.1890/06-1976.1. Hdl:1903/7873

- Futuyma, Douglas J.; Kirkpatrick, Mark (2017). "Mutation and variation". Evolution (Fourth ed.). Sunderland, Massachusetts: Sinauer Associates, Inc. Pp. 79–102. ISBN 978-1-60535-605-1

- Orr, H. Allen (February 2005). "The genetic theory of adaptation: a brief history". Nature Reviews Genetics. 6 (2): 119–127. Doi:10.1038/nrg1523. PMID 15716908

- Leroi, Armand Marie (2015). The Lagoon: How Aristotle Invented Science. Bloomsbury. Pp. 91–92, 273, 288. ISBN 978-1408836224

- Cullen, Esther (April 2008). "Adaptations in the kittiwake to cliff-nesting". Ibis. 99 (2): 275–302. Doi:10.1111/j.1474-919x.1957.tb01950.x

- Conway Morris, Simon (2005). Life's solution: inevitable humans in a lonely universe. Cambridge University Press. Pp. 164, 167, 170 and 235. Doi:10.2277/0521827043. ISBN 978-0-521-60325-6. OCLC 156902715

- Hastings, P J; Lupski, JR; Rosenberg, SM; Ira, G (2009). "Mechanisms of change in gene copy number". Nature Reviews Genetics. 10 (8): 551–564. Doi:10.1038/nrg2593. PMC 2864001. PMID 19597530

- Edited by Scott, Eugenie C.; Branch, Glenn (2006). Not in our classrooms : why intelligent design is wrong for our schools (1st ed.). Boston: Beacon Press. P. 47. ISBN 978-0807032787

Threats to Biodiversity

Factors such as pollution, overexploitation and climate change can cause loss of biodiversity. The species which are adversely affected by these factors are categorized as threatened, vulnerable, endangered, critically endangered and extinct in the wild. These causes of biodiversity loss as well as the different classifications of threatened species have been thoroughly discussed in this chapter.

The threats to biodiversity can be summarized in the following main points:

- Alteration and loss of the habitats: the transformation of the natural areas determines not only the loss of the vegetable species, but also a decrease in the animal species associated to them.

- Introduction of exotic species and genetically modified organisms: species originating from a particular area, introduced into new natural environments can lead to different forms of imbalance in the ecological equilibrium.

- Pollution: human activity influences the natural environment producing negative, direct or indirect, effects that alter the flow of energy, the chemical and physical constitution of the environment and abundance of the species.

- Climate change: for example, heating of the Earth's surface affects biodiversity because it endangers all the species that adapted to the cold due to the latitude (the Polar species) or the altitude (mountain species).

- Overexploitation of resources: when the activities connected with capturing and harvesting (hunting, fishing, farming) a renewable natural resource in a particular area is excessively intense, the resource itself may become exhausted, as for example, is the case of sardines, herrings, cod, tuna and many other species that man captures without leaving enough time for the organisms to reproduce.

THE MAJOR CAUSES FOR BIODIVERSITY LOSS

Loss of biodiversity occurs when either the habitat essential for the survival of a species is destroyed, or particular species are destroyed. The former is more common as habitat destruction is a fallout of development. The latter reason is encountered when particular species are exploited for economical gain or hunted for sport or food.

Extinction of species may also be due to environmental factors like ecological substitutions, biological factors and pathological causes which can be caused by nature or man.

Natural Causes for the Loss of Biodiversity

Natural causes include floods, earthquakes, landslides, natural competition between species, lack of pollination and diseases.

Man-Made Causes for the Loss of Biodiversity

- Destruction of habitat in the wake of developmental activities like housing, agriculture, construction of dams, reservoirs, roads, railway tracks, etc.

- Pollution, a gift of the industrial revolution can be given the pride of place for driving a variety of species in air, water and land towards extinction.

- Motorcars, air-conditioners and refrigerators, the three symbols of a modern, affluent society, have been instrumental in global warming and ozone depletion. They have drastically altered the climate with disastrous effects on the various species. Factories and power stations spewing out poisonous gases and effluents have fouled up the environment bringing death and disease to many species. Oil spills and discharge of sewage have ravaged the oceans and coastal habitats.

- A large number of species are threatened by overhunting, poaching and illegal trade.

- Indiscriminate use of toxic chemicals and pesticides and overexploitation of wildlife resources for commercial purposes are responsible for the rapid decline in the number of some species. The tiger for instance is hunted for its claws and other parts believed to be effective cures for various ailments of man. Snakes and crocodiles are killed in large numbers for their skin and minks, sable, ermine, etc., are in demand for the luxury and warmth of their fur.

- Genetic erosion arises from the loss (due to commercial and anthropogenic pressures) of habitats rich in biodiversity and from the disappearance of the traditional conservation practices of wild species in their habitats by rural and tribal people.

TYPES OF SPECIES THREATENED

Threatened Species

Threatened species are any species (including animals, plants, fungi, etc.) which are vulnerable to endangerment in the near future. Species that are threatened are sometimes characterised by the population dynamics measure of critical depensation, a mathematical measure of biomass related to population growth rate. This quantitative metric is one method of evaluating the degree of endangerment.

Vulnerable Species

A vulnerable species is one which has been categorized by the International Union for Conservation of Nature as likely to become endangered unless the circumstances that are threatening its survival and reproduction improve.

Vulnerability is mainly caused by habitat loss or destruction of the species home. Vulnerable habitat or species are monitored and can become increasingly threatened. Some species listed as "vulnerable" may be common in captivity, an example being the military macaw.

There are currently 5196 animals and 6789 plants classified as vulnerable, compared with 1998 levels of 2815 and 3222, respectively. Practices such as Cryoconservation of animal genetic resources have been enforced in efforts to conserve vulnerable breeds of livestock specifically.

Criteria

The International Union for Conservation of Nature uses several criteria to enter species in this category. A taxon is Vulnerable when it is not critically endangered or Endangered but is facing a high risk of extinction in the wild in the medium-term future, as defined by any of the following criteria (A to E):

A) Population reduction in the form of either of the following:

1. An observed, estimated, inferred or suspected population size reduction of $\geq 50\%$ over the last 10 years or three generations, whichever is the longer, provided the causes of the reduction are clearly reversible and understood and ceased. This measurement is based on (and specifying) any of the following:

    - Direct observation

    - An index of abundance appropriate for the taxon

    - A decline in area of occupancy, extent of occurrence or quality of habitat

    - Actual or potential levels of exploitation

    - The effects of introduced taxa, hybridisation, pathogens, pollutants, competitors or parasites

2. A reduction of at least 20%, projected or suspected to be met within the next ten years or three generations, whichever is the longer.

B) Extent of occurrence estimated to be less than 20,000 km^2 or area of occupancy estimated to be less than 2000 km^2, and estimates indicating any two of the following:

1. Severely fragmented or known to exist at no more than ten locations.

2. Continuing decline, inferred, observed or projected, in any of the following:

    - Extent of occurrence

    - Area of occupancy

    - Area, extent or quality of habitat

    - Number of locations or subpopulations

    - Number of mature individuals

3. Extreme fluctuations in any of the following:

   - Extent of occurrence

   - Area of occupancy

   - Number of locations or subpopulations

   - Number of mature individuals

C) Population estimated to number fewer than 10,000 mature individuals and either:

1. An estimated continuing decline of at least 10% within 10 years or three generations, whichever is longer, or

2. A continuing decline, observed, projected, or inferred, in numbers of mature individuals and population structure in the form of either:

   - Severely fragmented (i.e. no subpopulation estimated to contain more than 1000 mature individuals)

   - All mature individuals are in a single subpopulation

D) Population very small or restricted in the form of either of the following:

1. Population estimated to number less than 1000 mature individuals.

2. Population is characterised by an acute restriction in its area of occupancy (typically less than 20 km²) or in the number of locations (typically less than five). Such a taxon would thus be prone to the effects of human activities (or stochastic events whose impact is increased by human activities) within a very short period of time in an unforeseeable future, and is thus capable of becoming critically Endangered or even Extinct in a very short period.

E) Quantitative analysis showing the probability of extinction in the wild is at least 10% within 100 years.

The examples of vulnerable animal species are hyacinth macaw, mountain zebra, gaur, kea, black crowned crane and blue crane.

Endangered Species

An endangered species is a species which has been categorized as very likely to become extinct in the near future. Endangered (EN), as categorized by the International Union for Conservation of Nature (IUCN) Red List, is the second most severe conservation status for wild populations in the IUCN's schema after Critically Endangered (CR).

In 2012, the IUCN Red List featured 3,079 animal and 2,655 plant species as endangered (EN) worldwide. The figures for 1998 were, respectively, 1,102 and 1,197.

Many nations have laws that protect conservation-reliant species: for example, forbidding hunting, restricting land development or creating protected areas. Population numbers, trends and species' conservation status can be found at the lists of organisms by population.

Conservation Status

The conservation status of a species indicates the likelihood that it will become extinct. Many factors are considered when assessing the status of a species; e.g., such statistics as the number remaining, the overall increase or decrease in the population over time, breeding success rates, or known threats. The IUCN Red List of Threatened Species is the best-known worldwide conservation status listing and ranking system.

Over 50% of the world's species are estimated to be at risk of extinction. Internationally, 199 countries have signed an accord to create Biodiversity Action Plans that will protect endangered and other threatened species. In the United States, such plans are usually called Species Recovery Plans.

IUCN Red List

Though labelled a list, the IUCN Red List is a system of assessing the global conservation status of species that includes "Data Deficient" (DD) species – species for which more data and assessment is required before their status may be determined – as well species comprehensively assessed by the IUCN's species assessment process. Those species of "Near Threatened" (NT) and "Least Concern" (LC) status have been assessed and found to have relatively robust and healthy populations, though these may be in decline. Unlike their more general use elsewhere, the List uses the terms "endangered species" and "threatened species" with particular meanings: "Endangered" (EN) species lie between "Vulnerable" (VU) and "Critically Endangered" (CR) species, while "Threatened" species are those species determined to be Vulnerable, Endangered or Critically Endangered.

The IUCN categories, with examples of animals classified by them, include:

Extinct (EX)

No remaining individuals of the species.

Examples:

- Aurochs
- Ara atwoodi
- Blackfin cisco
- Caribbean monk seal
- Caspian tiger
- Dodo
- Eastern cougar

- Great auk
- Guam flycatcher
- Javan tiger
- Labrador duck
- Lesser bilby
- New Zealand quail
- Passenger pigeon

- Schomburgk's deer
- Steller's sea cow
- Thylacine
- Toolache wallaby
- California Grizzly Bear

Extinct in the Wild (EW)

Captive individuals survive, but there is no free-living, natural population.

Examples:

- Guam kingfisher
- Guam Rail
- Hawaiian crow
- Père David's deer
- Scimitar oryx
- Socorro dove
- South China tiger
- Wyoming toad

Critically Endangered (CR)

Faces an extremely high risk of extinction in the immediate future.

Examples:

- Addax
- African wild ass
- Alabama cavefish
- Amur leopard
- Javan rhino
- Arabian leopard
- Arakan forest turtle
- Asiatic cheetah
- Axolotl
- Wild Bactrian camel
- Black rhino
- Blue-throated macaw
- Brazilian merganser
- Brown spider monkey
- California condor
- Chinese alligator
- Chinese giant salamander
- Cross River gorilla
- Florida panther
- Gharial
- Hawaiian monk seal
- Hawksbill sea turtle
- Imperial woodpecker
- Ivory-billed Woodpecker
- Tristan albatross
- Amsterdam albatross
- Leadbeater's possum
- Mediterranean monk seal
- Northwest African cheetah
- Northern hairy-nosed wombat
- Philippine crocodile
- Red wolf

- Saiga
- Siamese crocodile
- Red-throated lorikeet
- Spix's macaw
- Southern bluefin tuna
- Rück's blue flycatcher
- Sumatran orangutan

- Sumatran rhinoceros
- Blue-fronted lorikeet
- Vaquita
- Yangtze river dolphin
- Western lowland gorilla
- Hawksbill sea turtle
- Kemp's ridley sea turtle

Endangered (EN)

Faces a high risk of extinction in the near future.

Examples:

- Mexican Wolf
- African penguin
- African wild dog
- Amur tiger
- Asian elephant
- Bengal tiger
- Australasian bittern
- Blue whale
- Bonobo
- Bornean orangutan
- Common chimpanzee
- Dhole
- Ethiopian wolf
- Flores crow
- Hispid hare
- Giant otter
- Goliath frog
- Grey parrot
- Green sea turtle
- Loggerhead sea turtle

- Grevy's zebra
- Humblot's heron
- Iberian lynx
- Indian pangolin
- Japanese crane
- Japanese night heron
- Lear's macaw
- Malayan tapir
- Markhor
- Malagasy pond heron
- Mountain gorilla
- Yellow headed amazon
- Purple-faced langur
- Red-breasted goose
- Rothschild's giraffe
- South Andean deer
- Anoa
- Takhi
- Toque macaque
- Vietnamese pheasant

- Volcano rabbit
- Wild water buffalo
- Whale shark
- White-eared night heron

- Whooping crane
- Tasmanian devil
- Red panda

Vulnerable (VU)

Faces a high risk of endangerment in the medium term.

Examples:

- Military macaw
- African leopard
- American paddlefish
- Common carp
- Clouded leopard
- Cheetah
- Dugong
- Far Eastern curlew
- Fossa
- Galapagos tortoise
- Gaur
- Blue headed macaw
- Blue-eyed cockatoo
- Golden hamster
- Great slaty woodpecker
- Hyacinth macaw

- Humboldt penguin
- Blue crane
- Lesser white-fronted goose
- Mandrill
- Maned sloth
- Montserrat oriole
- Mountain zebra
- Hawaiian goose
- Pacific walrus
- Sloth bear
- Snow leopard
- Takin
- Yak
- Great white shark
- American crocodile
- White-necked crow
- Dingo

Near-Threatened (NT)

May be considered threatened in the near future.

Examples:

- American bison
- Asian golden cat

- Blue-billed duck
- Emperor goose

- Emperor penguin
- Eurasian curlew
- jaguar
- Larch Mountain salamander
- Lesser long-nosed bat
- Magellanic penguin
- Maned wolf
- Margay

- montane solitary eagle
- Pampas cat
- Pallas's cat
- Reddish egret
- White rhinoceros
- Striped hyena
- Tiger shark
- White eared pheasant

Least Concern (LC)

No immediate threat to species' survival.

Examples:

- Black-bellied whistling duck
- Saltwater crocodile
- Indian peafowl
- Olive baboon
- Bald eagle
- Lesser bird of paradise
- Brown bear
- Brown rat
- Brown-throated sloth
- Canada goose
- Cane toad
- Common wood pigeon
- Magpie goose
- Grey wolf
- House mouse

- Wolverine
- Palm cockatoo
- Louisiana black bear
- Mallard
- Mute swan
- Eurasian magpie
- Red-billed quelea
- Common hill myna
- Red-tailed hawk
- Rock pigeon
- Blue and yellow macaw
- Southern elephant seal
- Freshwater crocodile
- Humpback whale
- Red howler monkey

Criteria for 'Endangered (EN)'

A) Reduction in population size based on any of the following:

1. An observed, estimated, inferred or suspected population size reduction of ≥ 70% over the last

10 years or three generations, whichever is the longer, where the causes of the reduction are clearly reversible and understood and ceased, based on (and specifying) any of the following:

- Direct observation.

- An index of abundance appropriate for the taxon.

- A decline in area of occupancy, extent of occurrence or quality of habitat.

- Actual or potential levels of exploitation.

- The effects of introduced taxa, hybridisation, pathogens, pollutants, competitors or parasites.

2. An observed, estimated, inferred or suspected population size reduction of ≥ 50% over the last 10 years or three generations, whichever is the longer, where the reduction or its causes may not have ceased or may not be understood or may not be reversible, based on (and specifying) any of (a) to (e) under A1.

3. A population size reduction of ≥ 50%, projected or suspected to be met within the next 10 years or three generations, whichever is the longer (up to a maximum of 100 years), based on (and specifying) any of (b) to (e) under A1.

4. An observed, estimated, inferred, projected or suspected population size reduction of ≥ 50% over any 10 year or three generation period, whichever is longer (up to a maximum of 100 years in the future), where the time period must include both the past and the future, and where the reduction or its causes may not have ceased or may not be understood or may not be reversible, based on (and specifying) any of (a) to (e) under A1.

B) Geographic range in the form of either B1 (extent of occurrence) or B2 (area of occupancy) or both:

1. Extent of occurrence estimated to be less than 5,000 km², and estimates indicating at least two of a-c:

- Severely fragmented or known to exist at no more than five locations.

- Continuing decline, inferred, observed or projected, in any of the following:
  ◦ Extent of occurrence
  ◦ Area of occupancy
  ◦ Area, extent or quality of habitat
  ◦ Number of locations or subpopulations
  ◦ Number of mature individuals

- Extreme fluctuations in any of the following:
  ◦ Extent of occurrence
  ◦ Area of occupancy
  ◦ Number of locations or subpopulations
  ◦ Number of mature individuals

2. Area of occupancy estimated to be less than 500 km², and estimates indicating at least two of a- c:

- Severely fragmented or known to exist at no more than five locations.

- Continuing decline, inferred, observed or projected, in any of the following:

    ° Extent of occurrence

    ° Area of occupancy

    ° Area, extent or quality of habitat

    ° Number of locations or subpopulations

    ° Number of mature individuals

- Extreme fluctuations in any of the following:

    ° Extent of occurrence

    ° Area of occupancy

    ° Number of locations or subpopulations

    ° Number of mature individuals

C) Population estimated to number fewer than 2,500 mature individuals and either:

1. An estimated continuing decline of at least 20% within five years or two generations, which-ever is longer, (up to a maximum of 100 years in the future).

2. A continuing decline, observed, projected, or inferred, in numbers of mature individuals and at least one of the follow (a-b):

- Population structure in the form of one of the following:

    ° No subpopulation estimated to contain more than 250 mature individuals

    ° At least 95% of mature individuals in one subpopulation

- Extreme fluctuations in number of mature individuals

D) Population size estimated to number fewer than 250 mature individuals.

E) Quantitative analysis showing the probability of extinction in the wild is at least 20% within 20 years or five generations, whichever is the longer (up to a maximum of 100 years).

- Near-critically endangered

- Particularly sensitive to poaching levels

- Near-endangered due to poaching

- May vary according to levels of tourism

Endangered Species in the United States

There is data from the United States that shows a correlation between human populations and threatened and endangered species. Using species data from the Database on the Economics and Management of Endangered Species (DEMES) database and the period that the Endangered Species Act (ESA) has been in existence, 1970 to 1997, a table was created that suggests a positive relationship between human activity and species endangerment.

Under the Endangered Species Act of 1973 in the United States, species may be listed as "endangered" or "threatened". The Salt Creek tiger beetle (Cicindela nevadica lincolniana) is an example of an endangered subspecies protected under the ESA. The US Fish and Wildlife Service as well as the National Marine Fisheries Service are held responsible for classifying and protecting endangered species, and adding a particular species to the list can be a long, controversial process.

Some endangered species laws are controversial. Typical areas of controversy include: criteria for placing a species on the endangered species list and criteria for removing a species from the list once its population has recovered; whether restrictions on land development constitute a "taking" of land by the government; the related question of whether private landowners should be compensated for the loss of uses of their lands; and obtaining reasonable exceptions to protection laws. Also lobbying from hunters and various industries like the petroleum industry, construction industry, and logging, has been an obstacle in establishing endangered species laws.

The Bush administration lifted a policy that required federal officials to consult a wildlife expert before taking actions that could damage endangered species. Under the Obama administration, this policy has been reinstated.

Being listed as an endangered species can have negative effect since it could make a species more desirable for collectors and poachers. This effect is potentially reducible, such as in China where commercially farmed turtles may be reducing some of the pressure to poach endangered species.

Another problem with the listing species is its effect of inciting the use of the "shoot, shovel, and shut-up" method of clearing endangered species from an area of land. Some landowners currently may perceive a diminution in value for their land after finding an endangered animal on it. They have allegedly opted to silently kill and bury the animals or destroy habitat, thus removing the problem from their land, but at the same time further reducing the population of an endangered species. The effectiveness of the Endangered Species Act – which coined the term "endangered species" – has been questioned by business advocacy groups and their publications but is nevertheless widely recognized by wildlife scientists who work with the species as an effective recovery tool. Nineteen species have been delisted and recovered and 93% of listed species in the northeastern United States have a recovering or stable population.

Currently, 1,556 known species in the world have been identified as near extinction or endangered and are under protection by government law. This approximation, however, does not take into consideration the number of species threatened with endangerment that are not included under the protection of such laws as the Endangered Species Act. According to NatureServe's global conservation status, approximately thirteen percent of vertebrates (excluding marine fish), seventeen percent of vascular plants, and six to eighteen percent of fungi are considered imperiled. Thus, in total, between seven and eighteen percent of the United States' known animals, fungi and plants are near extinction:416 This total is substantially more than the number of species protected in the United States under the Endangered Species Act.

Ever since mankind began hunting to preserve itself, over-hunting and fishing has been a large and dangerous problem. Of all the species who became extinct due to interference from mankind, the dodo, passenger pigeon, great auk, Tasmanian tiger and Steller's sea cow are some of the more well known examples; with the bald eagle, grizzly bear, American bison, Eastern timber wolf and sea turtle having been hunted to near-extinction. Many began as food sources seen as necessary for survival but became the target of sport. However, due to major efforts to prevent extinction, the bald eagle, or Haliaeetus leucocephalus is now under the category of Least Concern on the red list. A present-day example of the over-hunting of a species can be seen in the oceans as populations of certain whales have been greatly reduced. Large whales like the blue whale, bowhead whale, finback whale, gray whale, sperm whale and humpback whale are some of the eight whales which are currently still included on the Endangered Species List. Actions have been taken to attempt reduction in whaling and increase population sizes, including prohibiting all whaling in United States waters, the formation of the CITES treaty which protects all whales, along with the formation of the International Whaling Commission (IWC). But even though all of these movements have been put in place, countries such as Japan continue to hunt and harvest whales under the claim of "scientific purposes". Over-hunting, climatic change and habitat loss leads in landing species in endangered species list and could mean that extinction rates could increase to a large extent in the future.

Invasive Species

The introduction of non-indigenous species to an area can disrupt the ecosystem to such an extent that native species become endangered. Such introductions may be termed alien or invasive species. In some cases the invasive species compete with the native species for food or prey on the natives. In other cases a stable ecological balance may be upset by predation or other causes leading to unexpected species decline. New species may also carry diseases to which the native species have no resistance.

Critically Endangered

A critically endangered (CR) species is one that has been categorized by the International Union for Conservation of Nature (IUCN) as facing an extremely high risk of extinction in the wild.

As of 2014, there are 2,464 animal and 2,104 plant species with this assessment.

As the IUCN Red List does not consider a species extinct until extensive, targeted surveys have been conducted, species that are possibly extinct are still listed as critically endangered. IUCN maintains a list of "possibly extinct" CR(PE) and "possibly extinct in the wild" CR(PEW) species, modelled on categories used by BirdLife International to categorize these taxa.

International Union for Conservation of Nature Definition

To be defined as critically endangered in the Red List, a species must meet any of the following criteria (A–E) ("3G/10Y" signifies three generations or ten years—whichever is longer—over a maximum of 100 years; "MI" signifies Mature Individuals).

A: Population size reduction:

1. If the reasons for population reduction no longer occur and can be reversed, the population needs to have been reduced by at least 90%.

2. 3. and 4. If not, then the population needs to have been reduced by at least 80%.

B: Occurring over less than 100 km² OR the area of occupancy is less than 10 km²:

1. Severe habitat fragmentation or existing at just one location.

2. Decline in extent of occurrence, area of occupancy, area/extent/quality of habitat, number of locations/subpopulations, or amount of MI.

3. Extreme fluctuations in extent of occurrence, area of occupancy, number of locations/sub-populations, or amount of MI.

C: Declining population of less than 250 MI and either:

1. A decline of 25% over 3G/10Y;

2. Extreme fluctuations, or over 90% of MI in a single subpopulation, or no more than 50 MI in any one subpopulation.

D: Numbers less than 50 MI.

E: At least 50% chance of going Extinct in the Wild over 3G/10Y.

Extinct in the Wild

A species that is extinct in the wild (EW) is one that has been categorized by the International Union for Conservation of Nature as known only by living members kept in captivity or as a naturalized population outside its historic range due to massive habitat loss.

Examples of species and subspecies that are extinct in the wild include:

- Alagoas curassow
- Beloribitsa
- Black soft-shell turtle
- Cachorrito de charco palmal
- Escarpment cycad
- Franklinia
- Golden skiffia
- Guam kingfisher
- Guam rail

- Hawaiian crow or alalā (released into the wild; if the group maintains long term viability the IUCN may downgrade from extinct in the wild)

- Kihansi spray toad

- Oahu deceptor bush cricket

- Pedder galaxias

- Père David's deer

- Scimitar oryx

- Socorro dove

- Socorro isopod

- South China tiger

- Spix's macaw

- Wyoming toad

The Pinta Island tortoise (Geochelone nigra abingdoni) had only one living individual, named Lonesome George, until his death in June 2012. The tortoise was believed to be extinct in the mid-20th century, until Hungarian malacologist József Vágvölgyi spotted Lonesome George on the Galapagos island of Pinta on 1 December 1971. Since then, Lonesome George has been a powerful symbol for conservation efforts in general and for the Galapagos Islands in particular. With his death on 24 June 2012, the subspecies is again believed to be extinct. With the discovery of 17 hybrid Pinta tortoises located at nearby Wolf Volcano a plan has been made to attempt to breed the subspecies back into a pure state.

Not all species that are extinct in the wild are rare. For example, Ameca splendens, though extinct in the wild, was a popular fish among aquarists for some time, but hobbyist stocks have declined quite a lot more recently, placing its survival in jeopardy. However, the ultimate purpose of preserving biodiversity is to maintain ecological function. When a species exists only in captivity, it is ecologically extinct.

Reintroduction

Reintroduction is the deliberate release of species into the wild, from captivity or relocated from other areas where the species survives. This may be an option for certain species that are endangered or extinct in the wild. However, it may be difficult to reintroduce EW species into the wild, even if their natural habitats were restored, because survival techniques, which are often passed from parents to offspring during parenting, may be lost. While conservation efforts may preserve some of the genetics of a species, the species may never fully recover due to the loss of the natural memetics of the species.

An example of a successful reintroduction of an EW species is Przewalski's horse, which as of 2018 is considered to be an Endangered species, following reintroduction started in the 1990s.

IUCN RED LIST

IUCN Red List of Threatened Species, also called IUCN Red List, one of the most well-known objective assessment systems for classifying the status of plants, animals, and other organisms threatened with extinction. The International Union for Conservation of Nature (IUCN) unveiled this assessment system in 1994. It contains explicit criteria and categories to classify the conservation status of individual species on the basis of their probability of extinction.

The IUCN system uses a set of five quantitative criteria to assess the extinction risk of a given species. In general, these criteria consider:

1. The rate of population decline.

2. The geographic range.

3. Whether the species already possesses a small population size.

4. Whether the species is very small or lives in a restricted area.

5. Whether the results of a quantitative analysis indicate a high probability of extinction in the wild.

After a given species has been thoroughly evaluated, it is placed into one of several categories. (The details of each have been condensed to highlight two or three of the category's most salient points below.) In addition, three of the categories (CR, EN, and VU) are contained within the broader notion of "threatened." The IUCN Red List of Threatened Species recognizes several categories of species status:

1. Extinct (EX), a designation applied to species in which the last individual has died or where systematic and time-appropriate surveys have been unable to log even a single individual.

2. Extinct in the Wild (EW), a category containing those species whose members survive only in captivity or as artificially supported populations far outside their historical geographic range.

3. Critically Endangered (CR), a category containing those species that possess an extremely high risk of extinction as a result of rapid population declines of 80 to more than 90 percent over the previous 10 years (or three generations), a current population size of fewer than 50 individuals, or other factors.

4. Endangered (EN), a designation applied to species that possess a very high risk of extinction as a result of rapid population declines of 50 to more than 70 percent over the previous 10 years (or three generations), a current population size of fewer than 250 individuals, or other factors.

5. Vulnerable (VU), a category containing those species that possess a very high risk of extinction as a result of rapid population declines of 30 to more than 50 percent over the previous 10 years (or three generations), a current population size of fewer than 1,000 individuals, or other factors.

6. Near Threatened (NT), a designation applied to species that are close to becoming threatened or may meet the criteria for threatened status in the near future.

7. Least Concern (LC), a category containing species that are pervasive and abundant after careful assessment.

8. Data Deficient (DD), a condition applied to species in which the amount of available data related to its risk of extinction is lacking in some way. Consequently, a complete assessment cannot be performed. Thus, unlike the other categories in this list, this category does not describe the conservation status of a species.

9. Not Evaluated (NE), a category used to include any of the nearly 1.9 million species described by science but not assessed by the IUCN.

All else being equal, a species experiencing an 90 percent decline over 10 years (or three generations), for example, would be classified as critically endangered. Likewise, another species undergoing a 50 percent decline over the same period would be classified as endangered, and one experiencing a 30 percent reduction over the same time frame would be considered vulnerable. It is important to understand, however, that a species cannot be classified by using one criterion alone; it is essential for the scientist doing the assessment to consider all five criteria when determining the status of the species.

Each year thousands of scientists around the world assess or reassess species. The IUCN Red List is subsequently updated with these new data once the assessments have been checked for accuracy. In this way, the information helps to provide a continual spotlight on the status of the world's at-risk plants, animals, and other organisms. As a result, interested parties, such as national governments and conservation organizations, may use the information provided in the IUCN Red List to prioritize their own species-protection efforts.

The IUCN Red List brings into focus the ongoing decline of Earth's biodiversity and the influence humans have on life on the planet. It provides a globally accepted standard with which to measure the conservation status of species over time. By 2019, 96,500 species had been assessed by using the IUCN Red List categories and criteria. Of these, more than 26,500 species of plants, animals, and others fall into the threatened categories (CR, EN, and VU). Today the list appears as an online database available to the public. Scientists can analyze the percentage of species in a given category and how these percentages change over time; they can also analyze the threats and conservation measures that underpin the observed trends.

IUCN Red List of Threatened Species

The IUCN Red List is a critical indicator of the health of the world's biodiversity. Far more than a list of species and their status, it is a powerful tool to inform and catalyze action for biodiversity conservation and policy change, critical to protecting the natural resources we need to survive. It provides information about range, population size, habitat and ecology, use and/or trade, threats, and conservation actions that will help inform necessary conservation decisions.

The IUCN Red List is used by government agencies, wildlife departments, conservation-related non-governmental organizations (NGOs), natural resource planners, educational organizations, students, and the business community. The Red List process has become a massive enterprise involving the IUCN Global Species Program staff, partner organizations and experts in the IUCN

Species Survival Commission and partner networks who compile the species information to make The IUCN Red List the indispensable product it is today.

To date, many species groups including mammals, amphibians, birds, reef building corals and conifers have been comprehensively assessed. As well as assessing newly recognized species, the IUCN Red List also re-assesses the status of some existing species, sometimes with positive stories to tell. For example, good news such as the downlisting (i.e. improvement) of a number of species on the IUCN Red List categories scale, due to conservation efforts. The bad news, however, is that biodiversity is declining. Currently there are more than 98,500 species on The IUCN Red List, and more than 27,000 are threatened with extinction, including 40% of amphibians, 34% of conifers, 33% of reef building corals, 25% of mammals and 14% of birds.

Despite the high proportions of threatened species, we are working to reverse, or at least halt, the decline in biodiversity. Increased assessments will help to build The IUCN Red List into a more complete 'Barometer of Life'. To do this we need to increase the number of species assessed to at least 160,000 by 2020. This will improve the global taxonomic coverage and thus provide a stronger base to enable better conservation and policy decisions. The IUCN Red List is crucial not only for helping to identify those species needing targeted recovery efforts, but also for focusing the conservation agenda by identifying the key sites and habitats that need to be protected. Ultimately, The IUCN Red List helps to guide and inform future conservation and funding priorities.

References

- Causes-of-the-loss-of-biodiversity, loss-of-biodiversity, biodiversity1, argomento: eniscuola.Net, retrieved 20 july, 2019

- Program, u.S. Fish and wildlife service/endangered species. "Endangered species program - laws & policies - endangered species act - section 3 definitions". Www.Fws.Gov. Archived from the original on 29 april 2017. Retrieved 8 may 2018

- Conservation-of-biodiversity, biodiversity-1, environment, energy: vikaspedia.In, retrieved 30 may, 2019

- "Red list overview". Iucn. February 2011. Archived from the original on may 27, 2012. Retrieved 2 june 2012

- Iucn-red-list-of-threatened-species: britannica.Com, retrieved 25 august, 2019

- Abramov, a.; Belant, j. & Wozencraft, c. (2009). "Gulo gulo". Iucn red list of threatened species. Version 2009.2. International union for conservation of nature. Retrieved 2010-01-25

- Donaldson, j.S. (2010). "Encephalartos brevifoliolatus". The iucn red list of threatened species. Iucn. 2010: E.T41882a10566751. Doi:10.2305/Iucn.Uk.2010-3.Rlts.T41882a10566751.En. Retrieved 14 january 2018

- Nicholls, h. (2006). Lonesome george: the life and loves of a conservation icon. London, england: macmillan science. Isbn 1-4039-4576-4

- Background-history: iucnredlist.Org, retrieved 3 july, 2019

Biodiversity Conservation

The management of natural resources with an aim to sustain biodiversity in ecosystems, species and the evolutionary processes is known as conservation of biodiversity. This chapter has been carefully written to provide an easy understanding of the importance of biodiversity and the diverse types of conservation such as ex situ conservation and in situ conservation.

IMPORTANCE OF BIODIVERSITY

Biodiversity boosts ecosystem productivity where each species, no matter how small, all have an important role to play.

For example,

- A larger number of plant species means a greater variety of crops.

- Greater species diversity ensures natural sustainability for all life forms.

- Healthy ecosystems can better withstand and recover from a variety of disasters.

And so, while we dominate this planet, we still need to preserve the diversity in wildlife.

A healthy biodiversity provides a number of natural services for everyone:

- Ecosystem services, such as:

    ◦ Protection of water resources.

    ◦ Soils formation and protection.

    ◦ Nutrient storage and recycling.

    ◦ Pollution breakdown and absorption.

    ◦ Contribution to climate stability.

    ◦ Maintenance of ecosystems.

    ◦ Recovery from unpredictable events.

- Biological resources, such as:

    ◦ Food.

- ◦ Medicinal resources and pharmaceutical drugs.

- ◦ Wood products.

- ◦ Ornamental plants.

- ◦ Breeding stocks, population reservoirs.

- ◦ Future resources.

- ◦ Diversity in genes, species and ecosystems.

- Social benefits, such as:

- ◦ Education and monitoring.

- ◦ Recreation and tourism.

- ◦ Cultural values.

The cost of replacing these (if possible) would be extremely expensive. It therefore makes economic and development sense to move towards sustainability.

A report from Nature magazine also explains that genetic diversity helps to prevent the chances of extinction in the wild (and claims to have shown proof of this).

To prevent the well known and well documented problems of genetic defects caused by in-breeding, species need a variety of genes to ensure successful survival. Without this, the chances of extinction increases.

And as we start destroying, reducing and isolating habitats, the chances for interaction from species with a large gene pool decreases.

Species Depend on Each Other

While there might be survival of the fittest within a given species, each species depends on the services provided by other species to ensure survival. It is a type of cooperation based on mutual survival and is often what a balanced ecosystem refers to.

Soil, Bacteria, Plants: The Nitrogen Cycle

The relationship between soil, plants, bacteria and other life is also referred to as the nitrogen cycle:

As an example, consider all the species of animals and organisms involved in a simple field used in agriculture. As summarized from Vandana Shiva, Stolen Harvest:

- Crop byproducts feed cattle.

- Cattle waste feeds the soil that nourish the crops.

- Crops, as well as yielding grain also yield straw.

- ◦ Straw provides organic matter and fodder.

- ○   Crops are therefore food sources for humans and animals
- Soil organisms also benefit from crops:
  - ○   Bacteria feed on the cellulose fibers of straw that farmers return to the soil.
  - ○   Amoebas feed on bacteria making lignite fibers available for uptake by plants.
  - ○   Algae provide organic matter and serve as natural nitrogen fixers.
  - ○   Rodents that bore under the fields aerate the soil and improve its water-holding capacity.
  - ○   Spiders, centipedes and insects grind organic matter from the surface soil and leave behind enriched droppings.
  - ○   Earthworms contribute to soil fertility.
  - ○   They provide aerage, drainage and maintain soil structure.
  - ○   According to Charles Darwin, It may be doubted whether there are many other animals which have played so important a part in the history of creatures.
  - ○   The earthworm is like a natural tractor, fertilizer factory and dam, combined.
- Industrial-farming techniques would deprive these diverse species of food sources and instead assault them with chemicals, destroying the rich biodiversity in the soil and with it the basis for the renewal of the soil fertility.

Shiva, a prominent Indian scientist and activist goes on to detail the costs associated with destroying this natural diversity and traditional farming techniques which recognize this, and replacing this with industrial processes which go against the nature of diversity sustainability.

Large Carnivores Essential for Healthy Ecosystems

Three quarters of the world's big carnivores are in decline. A study in the journal Science, notes that these large animals — such as lions, leopards, wolves and bears — are in decline, due to declining habitats and persecution by humans.

This also has a negative impact on the environment, perhaps partly formed by outdated-views that predators are harmful for other wildlife. As the study notes, human actions cannot fully replace the role of large carnivores because these large carnivores are an intrinsic part of an ecosystem's biodiversity.

As a simple example, the loss of a large carnivore may mean in the short term the herbivores they prey on may increase in numbers but this can also result in a deterioration of the environment as the herbivores can graze more, largely unchecked. Human intervention to perform the same services would be more costly.

Conservation biology matured in the mid-20th century as ecologists, naturalists and other scientists began to research and address issues pertaining to global biodiversity declines.

The conservation ethic advocates management of natural resources for the purpose of sustaining biodiversity in species, ecosystems, the evolutionary process and human culture and society.

Conservation biology is reforming around strategic plans to protect biodiversity. Preserving global biodiversity is a priority in strategic conservation plans that are designed to engage public policy and concerns affecting local, regional and global scales of communities, ecosystems and cultures. Action plans identify ways of sustaining human well-being, employing natural capital, market capital and ecosystem services.

Protection and Restoration Techniques

Removal of exotic species will allow the species that they have negatively impacted to recover their ecological niches. Exotic species that have become pests can be identified taxonomically (e.g., with Digital Automated Identification System (DAISY), using the barcode of life). Removal is practical only given large groups of individuals due to the economic cost.

- As sustainable populations of the remaining native species in an area become assured, "missing" species that are candidates for reintroduction can be identified using databases such as the Encyclopedia of Life and the Global Biodiversity Information Facility.

- Biodiversity banking places a monetary value on biodiversity. One example is the Australian Native Vegetation Management Framework.

- Gene banks are collections of specimens and genetic material. Some banks intend to reintroduce banked species to the ecosystem (e.g., via tree nurseries).

- Reduction and better targeting of pesticides allows more species to survive in agricultural and urbanized areas.

- Location-specific approaches may be less useful for protecting migratory species. One approach is to create wildlife corridors that correspond to the animals' movements. National and other boundaries can complicate corridor creation.

TYPES OF CONSERVATION

Ex Situ Conservation

Conserving biodiversity outside the areas where they naturally occur is known as ex situ conservation. Here, animals and plants are reared or cultivated in areas like zoological or botanical parks.

Reintroduction of an animal or plant into the habitat from where it has become extinct is another form of ex situ conservation. For example, the Gangetic gharial has been reintroduced in the rivers of Uttar Pradesh, Madhya Pradesh and Rajasthan where it had become extinct.

Seedbanks, botanical, horticultural and recreational gardens are important centres for ex situ conservation.

In Situ Conservation

Conserving the animals and plants in their natural habitats is known as in situ conservation. This includes the establishment of:

- National parks and sanctuaries

- Biosphere reserves

- Nature reserves

- Reserved and protected forests

- Preservation plots

- Reserved forests

Agrobiodiversity Conservation

After the introduction of cotton, tobacco, sugarcane, sunflower, soyabean and so on, farmers became victims of monocultures in their greed for money. Therefore many of the indigenous varieties of crops were lost. Moreover, the hybrid varieties of fruits and vegetables (e.g. tomatoes), introduced for pulp are more susceptible to disease and pests. Though hybrid varieties are preferred, traditional wild varieties of the seeds should be conserved for future use in the event of an epidemic which would completely wipe out the hybrids.

Botanical gardens, agricultural departments, seed banks etc., alone should not be given the responsibility of agrobiodiversity conservation. Every farmer, gardener and cultivator should be aware of his role in preserving and conserving agrobiodiversity.

PROTECTED AREAS

Protected areas or conservation areas are locations which receive protection because of their recognized natural, ecological or cultural values. There are several kinds of protected areas, which vary by level of protection depending on the enabling laws of each country or the regulations of the international organizations involved.

The term "protected area" also includes Marine Protected Areas, the boundaries of which will include some area of ocean, and Transboundary Protected Areas that overlap multiple countries which remove the borders inside the area for conservation and economic purposes. There are over 161,000 protected areas in the world (as of October 2010) with more added daily, representing between 10 and 15 percent of the world's land surface area. By contrast, only 1.17% of the world's oceans is included in the world's ~6,800 Marine Protected Areas.

Protected areas are essential for biodiversity conservation, often providing habitat and protection from hunting for threatened and endangered species. Protection helps maintain ecological processes that cannot survive in most intensely managed landscapes and seascapes.

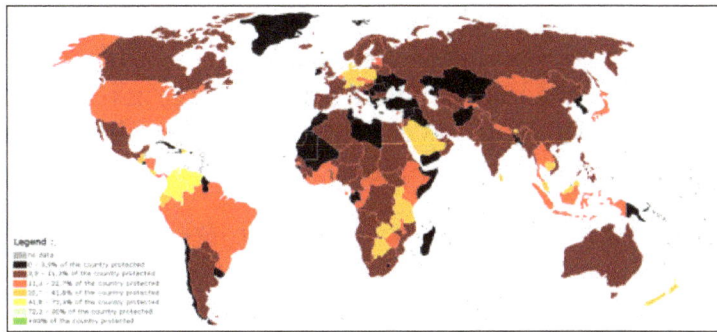

World map with total percentage of each country under protection.

Generally, protected areas are understood to be those in which human occupation or at least the exploitation of resources is limited. The definition that has been widely accepted across regional and global frameworks has been provided by the International Union for Conservation of Nature (IUCN) in its categorisation guidelines for protected areas. The definition is as follows:

"A clearly defined geographical space, recognized, dedicated and managed, through legal or other effective means, to achieve the long-term conservation of nature with associated ecosystem services and cultural values."

Protection of Natural Resources

The objective of protected areas is to conserve biodiversity and to provide a way for measuring the progress of such conservation. Protected areas will usually encompass several other zones that have been deemed important for particular conservation uses, such as Important Bird Areas (IBA) and Endemic Bird Areas (EBA), Centres of Plant Diversity (CPD), Indigenous and Community Conserved Areas (ICCA), Alliance for Zero Extinction Sites (AZE) and Key Biodiversity Areas (KBA) among others. Likewise, a protected area or an entire network of protected areas may lie within a larger geographic zone that is recognised as a terrestrial or marine ecoregions, or a crisis ecoregions for example. As a result, Protected Areas can encompass a broad range of governance types. Indeed, governance of protected areas has emerged a critical factor in their success.

Subsequently, the range of natural resources that any one protected area may guard is vast. Many will be allocated primarily for species conservation whether it be flora or fauna or the relationship between them, but protected areas are similarly important for conserving sites of (indigenous) cultural importance and considerable reserves of natural resources such as:

- Carbon stocks: Carbon emissions from deforestation account for an estimated 20% of global carbon emissions, so in protecting the worlds carbon stocks greenhouse gas emissions are reduced and longterm land cover change is prevented, which is an effective strategy in the struggle against global warming. Of all global terrestrial carbon stock, 15.2% is contained within protected areas. Protected areas in South America hold 27% of the world's carbon stock, which is the highest percentage of any country in both absolute terms and as a proportion of the total stock.

- Rainforests: 18.8% of the world's forest is covered by protected areas and sixteen of the twenty forest types have 10% or more protected area coverage. Of the 670 ecoregions with forest cover, 54% have 10% or more of their forest cover protected under IUCN Categories I – VI.

- Mountains: Nationally designated protected areas cover 14.3% of the world's mountain areas, and these mountainous protected areas made up 32.5% of the world's total terrestrial protected area coverage in 2009. Mountain protected area coverage has increased globally by 21% since 1990 and out of the 198 countries with mountain areas, 43.9% still have less than 10% of their mountain areas protected.

Annual updates on each of these analyses are made in order to make comparisons to the Millennium Development Goals and several other fields of analysis are expected to be introduced in the monitoring of protected areas management effectiveness, such as freshwater and marine or coastal studies which are currently underway, and islands and drylands which are currently in planning.

IUCN Protected Area Management Categories

Through its World Commission on Protected Areas (WCPA), the IUCN has developed six Protected Area Management Categories that define protected areas according to their management objectives, which are internationally recognised by various national governments and the United Nations. The categories provide international standards for defining protected areas and encourage conservation planning according to their management aims.

IUCN Protected Area Management Categories:

- Category Ia — Strict Nature Reserve

- Category Ib — Wilderness Area

- Category II — National Park

- Category III — Natural Monument or Feature

- Category IV — Habitat/Species Management Area

- Category V — Protected Landscape/Seascape

- Category VI – Protected Area with sustainable use of natural resources

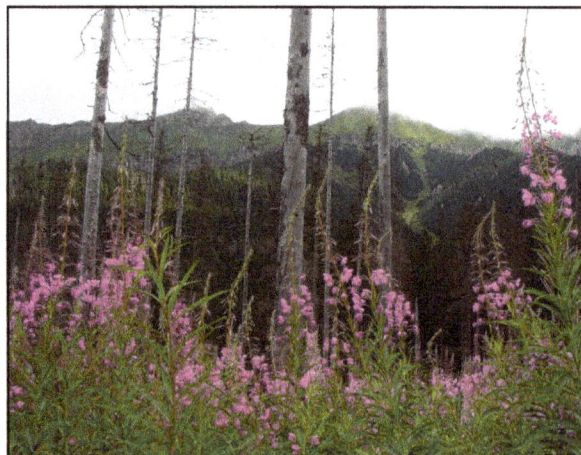

Strict Nature reserve Belianske Tatras in Slovakia.

Challenges

How to manage areas protected for conservation brings up a range of challenges - whether it be regarding the local population, specific ecosystems or the design of the reserve itself - and because of the many unpredicatable elements in ecology issues, each protected area requires a case-specific set of guidelines.

Enforcing protected area boundaries is a costly and labour-heavy endeavour, particularly if the allocation of a new protected region places new restrictions on the use of resources by the native people which may lead to their subsequent displacement. This has troubled relationships between conservationists and rural communities in many protected regions and is often why many Wildlife Reserves and National Parks face the human threat of poaching for the illegal bushmeat or trophy trades, which are resorted to as an alternative form of substinence.

There is increasing pressure to take proper account of human needs when setting up protected areas and these sometimes have to be "traded off" against conservation needs. Whereas in the past governments often made decisions about protected areas and informed local people afterwards, today the emphasis is shifting towards greater discussions with stakeholders and joint decisions about how such lands should be set aside and managed. Such negotiations are never easy but usually produce stronger and longer-lasting results for both conservation and people.

In some countries, protected areas can be assigned without the infrastructure and networking needed to substitute consumable resources and subtantiatively protect the area from development or misuse. The soliciting of protected areas may require regulation to the level of meeting demands for food, feed, livestock and fuel, and the legal enforcement of not only the protected area itself but also 'buffer zones' surrounding it, which may help to resist destabilisation.

Schweizerischer National Park in the Swiss Alps is a Strict Nature Reserve (Category Ia).

The Jaldapara National Park in West Bengal, India is a Habitat Management Area (Category IV).

Effectiveness

One of the main concerns regarding protected areas on land and sea is their effectiveness at preventing the ongoing loss of biodiversity. There are multiple case studies indicating the positive effects of protected areas on terrestrial and marine species. However, those cases do not represent the majority of protected areas. Several limitations that may preclude their success include: their small size and large isolation to each other (both of these factors influence the maintenance of species), their limited role at preventing the many factors affecting biodiversity (e.g. climate change, invasive species, pollution), their large cost and their increasing conflict with human demands for nature's goods and services.

National Parks

A national park is a park in use for conservation purposes. Often it is a reserve of natural, semi-natural, or developed land that a sovereign state declares or owns. Although individual nations designate their own national parks differently, there is a common idea: the conservation of 'wild nature' for posterity and as a symbol of national pride.

An international organization, the International Union for Conservation of Nature (IUCN), and its World Commission on Protected Areas (WCPA), has defined "National Park" as its Category II type of protected areas. According to the IUCN, 6,555 national parks worldwide met its criteria in 2006. IUCN is still discussing the parameters of defining a national park.

While this type of national park had been proposed previously, the United States established the first "public park or pleasuring-ground for the benefit and enjoyment of the people", Yellowstone National Park, in 1872. Although Yellowstone was not officially termed a "national park" in its establishing law, it was always termed such in practice and is widely held to be the first and oldest national park in the world. However, the Tobago Main Ridge Forest Reserve, and the area surrounding Bogd Khan Uul Mountain are seen as the oldest legally protected areas, predating Yellowstone by nearly a century.

Bogd Khan Uul National Park, Mongolia. One of the
earliest preserved areas now called a national park.

In 1969, the IUCN declared a national park to be a relatively large area with the following defining characteristics:

- One or several ecosystems not materially altered by human exploitation and occupation, where plant and animal species, geomorphological sites and habitats are of special scientific, educational, and recreational interest or which contain a natural landscape of great beauty.

- Highest competent authority of the country has taken steps to prevent or eliminate exploitation or occupation as soon as possible in the whole area and to effectively enforce the respect of ecological, geomorphological, or aesthetic features which have led to its establishment.

- Visitors are allowed to enter, under special conditions, for inspirational, educative, cultural, and recreative purposes.

In 1971, these criteria were further expanded upon leading to more clear and defined benchmarks to evaluate a national park. These include:

- Minimum size of 1,000 hectares within zones in which protection of nature takes precedence.

- Statutory legal protection.

- Budget and staff sufficient to provide sufficient effective protection.

- Prohibition of exploitation of natural resources (including the development of dams) qualified by such activities as sport, hunting, fishing, the need for management, facilities, etc.

While the term national park is now defined by the IUCN, many protected areas in many countries are called national park even when they correspond to other categories of the IUCN Protected Area Management Definition, for example:

- Swiss National Park, Switzerland: IUCN Ia – Strict Nature Reserve.

- Everglades National Park, United States: IUCN Ib – Wilderness Area.

- Victoria Falls National Park, Zimbabwe: IUCN III – National Monument.

- Vitosha National Park, Bulgaria: IUCN IV – Habitat Management Area.

- New Forest National Park, United Kingdom: IUCN V – Protected Landscape.

- Etniko Ygrotopiko Parko Delta Evrou, Greece: IUCN VI – Managed Resource Protected Area.

While national parks are generally understood to be administered by national governments (hence the name), in Australia national parks are run by state governments and predate the Federation of Australia; similarly, national parks in the Netherlands are administered by the provinces. In Canada, there are both national parks operated by the federal government and provincial or territorial parks operated by the provincial and territorial governments, although nearly all are still national parks by the IUCN definition.

In many countries, including Indonesia, the Netherlands, and the United Kingdom, national parks do not adhere to the IUCN definition, while some areas which adhere to the IUCN definition are not designated as national parks.

Notable Parks

The largest national park in the world meeting the IUCN definition is the Northeast Greenland National Park, which was established in 1974.

The smallest official national park in the world is Isles des Madeleines National Park. Its area of just .45 square kilometres (0.17 sq mi) was established as a national park in 1976.

Economic Ramifications

Countries with a large nature-based tourism industry, such as Costa Rica, often experience a huge economic effect on park management as well as the economy of the country as a whole.

Tourism

Tourism to national parks has increased considerably over time. In Costa Rica for example, a mega-diverse country, tourism to parks has increased by 400% from 1985 to 1999. The term national park is perceived as a brand name that is associated with nature-based tourism and it symbolizes a "high quality natural environment with a well-designed tourist infrastructure".

Staff

The duties of a park ranger are to supervise, manage, and/or perform work in the conservation and use of Federal park resources. This involves functions such as park conservation; natural, historical, and cultural resource management; and the development and operation of interpretive and recreational programs for the benefit of the visiting public. Park rangers also have fire fighting responsibilities and execute search and rescue missions. Activities also include heritage interpretation to disseminate information to visitors of general, historical, or scientific information. Management of resources such as wildlife, lakeshores, seashores, forests, historic buildings, battlefields, archeological properties, and recreation areas are also part of the job of a park ranger. Since the establishment of the National Park Service in the US in 1916, the role of the park ranger has shifted from merely being a custodian of natural resources to include several activities that are associated with law enforcement. They control traffic and investigate violations, complaints, trespass/encroachment, and accidents.

Wildlife Sanctuary

A wildlife refuge, also known as a wildlife sanctuary, is a naturally occurring sanctuary, such as an island, that provides protection for wildlife species from hunting, predation, competition or poaching; it is a protected area, a geographic landmark within which wildlife is protected. Refuges preserve animals that are endangered or even the animal which seems to go endangered in the future.

Such wildlife refuges are generally officially designated territories. They are created by government legislation, publicly or privately owned. Unofficial sanctuaries can also occur as a result of human accidents; the Chernobyl Exclusion Zone has in practice become a wildlife refuge since very few people live in the area. Wildlife has flourished in the Zone since the Chernobyl nuclear accident in 1986.

In the United States, the U.S. Fish and Wildlife Service applies the term "refuge" to various categories of areas administered by the Secretary of the Interior for the conservation of fish and wildlife. The Refuge System includes areas administered for the protection and conservation of fish and wildlife that are threatened with extinction, as well as wildlife ranges, game ranges, wildlife management areas, and waterfowl production areas.

The first North American wildlife refuge, Lake Merritt Wildlife scantuary at Lake Merritt, was established by Samuel Merritt and enacted in California state law in 1870 as the first government owned refuge. The first federally owned refuge in the United States is Pelican Island National Wildlife Refuge and was established by Theodore Roosevelt in 1903 as part of his Square Deal campaign to improve the country. At the time, setting aside land for wildlife was not a constitutional right of the president. More recently, a bi-partisan group of US House of Representatives members established the Congressional Wildlife Refuge Caucus to further the needs of the National Wildlife Refuge System in the US Congress.

Today there are several national and international organizations that have taken the responsibility of supervising numerous systems of non-profit animal sanctuaries and refuges in order to provide a general system for sanctuaries to follow. Among them, the American Sanctuary Association monitors and aids in various facilities to care for exotic wildlife. Their accredited facilities follow high standards and a rigid application processes to ensure that the animals under their care are avidly cared for and maintained. The number of sanctuaries has substantially increased over the past few years.

Forest Reserves

Forest reserves are portions of state lands where commercial harvesting of wood products is excluded in order to capture elements of biodiversity that can be missing from sustainably harvested sites. Small (patch) reserves will conserve sensitive, localized resources such as steep slopes, fragile soils, and habitat for certain rare species that benefit from intact forest canopies. Large (matrix) reserves will represent the diversity of relatively un-fragmented forest landscapes remaining in Massachusetts today. Matrix reserves will ultimately support a wider diversity of tree sizes and ages than typically occurs on sustainably harvested sites, and will also support structures and processes associated with extensive accumulations of large woody debris that are typically absent from harvested sites.

Matrix reserves will ultimately include a wide range of tree sizes and ages, from large, old trees 200-500 years old, to small, young trees that occur in open gaps where old trees have died or been blown over. Matrix reserves will ultimately feature extensive "pit and mound" micro-topography that occurs when old trees are blown over and their roots are pulled from the ground. Pits are formed when roots of large trees are pulled out of the ground during a natural disturbance like a wind storm. Pits collect moisture, organic matter, and nutrients over time, and provide unique, protected micro-climates for plants and invertebrate wildlife. Over time, the exposed roots of toppled trees degrade and form mounds characterized by extreme soil conditions of low moisture, low organic matter, and low nutrients that are markedly different from, yet in close proximity to pits originally occupied by the roots. The trunks and branches of large trees that are toppled during wind storms will accumulate as large woody debris in the forest, and will support decades or even centuries of activity by micro-organisms and invertebrate wildlife that occupy, feed upon, and ultimately break down these massive stores of organic material.

Zoological Parks

A zoo (also called an animal park or menagerie) is a facility in which all animals are housed within enclosures, displayed to the public, and in which they may also breed.

The abbreviation "zoo" was first used of the London Zoological Gardens, which was opened for scientific study in 1828 and to the public in 1857. In the United States alone, zoos are visited by over 181 million people annually.

Type

Zoo animals live in enclosures that often attempt to replicate their natural habitats or behavioral patterns, for the benefit of both the animals and visitors. Nocturnal animals are often housed in buildings with a reversed light-dark cycle, i.e. only dim white or red lights are on during the day so the animals are active during visitor hours, and brighter lights on at night when the animals sleep. Special climate conditions may be created for animals living in extreme environments, such as penguins. Special enclosures for birds, mammals, insects, reptiles, fish, and other aquatic life forms have also been developed. Some zoos have walk-through exhibits where visitors enter enclosures of non-aggressive species, such as lemurs, marmosets, birds, lizards, and turtles. Visitors are asked to keep to paths and avoid showing or eating foods that the animals might snatch.

Safari Park

Some zoos keep animals in larger, outdoor enclosures, confining them with moats and fences, rather than in cages. Safari parks, also known as zoo parks and lion farms, allow visitors to drive through them and come in close proximity to the animals. Sometimes, visitors are able to feed animals through the car windows. The first safari park was Whipsnade Park in Bedfordshire, England, opened by the Zoological Society of London in 1931 which today (2014) covers 600 acres (2.4 km²). Since the early 1970s, an 1,800 acre (7 km²) park in the San Pasqual Valley near San Diego has featured the San Diego Zoo Safari Park, run by the Zoological Society of San Diego. One of two state-supported zoo parks in North Carolina is the 2,000-acre (8.1 km2) North Carolina Zoo in Asheboro. The 500-acre (2.0 km2) Werribee Open Range Zoo in Melbourne, Australia, displays animals living in an artificial savannah.

Aquaria

The first public aquarium was opened at the London Zoo in 1853. This was followed by the opening of public aquaria in continental Europe (e.g. Paris in 1859, Hamburg in 1864, Berlin in 1869, and Brighton in 1872) and the United States (e.g. Boston in 1859, Washington in 1873, San Francisco Woodward's Garden in 1873, and the New York Aquarium at Battery Park in 1896).

Roadside Zoos

Roadside zoos are found throughout North America, particularly in remote locations. They are often small, for-profit zoos, often intended to attract visitors to some other facility, such as a gas station. The animals may be trained to perform tricks, and visitors are able to get closer to them than in larger zoos. Since they are sometimes less regulated, roadside zoos are often subject to accusations of neglect and cruelty.

In June 2014 the Animal Legal Defense Fund filed a lawsuit against the Iowa-based roadside Cricket Hollow Zoo for violating the Endangered Species Act by failing to provide proper care for its animals. Since filing the lawsuit, ALDF has obtained records from investigations conducted by the USDA's Animal and Plant Health Inspection Services; these records show that the zoo is also violating the Animal Welfare Act.

Petting Zoos

A petting zoo, also called petting farms or children's zoos, features a combination of domestic animals and wild species that are docile enough to touch and feed. To ensure the animals' health, the food is supplied by the zoo, either from vending machines or a kiosk nearby.

Animal Theme Parks

An animal theme park is a combination of an amusement park and a zoo, mainly for entertaining and commercial purposes. Marine mammal parks such as Sea World and Marineland are more elaborate dolphinariums keeping whales, and containing additional entertainment attractions. Another kind of animal theme park contains more entertainment and amusement elements than the classical zoo, such as a stage shows, roller coasters, and mythical creatures. Some examples are Busch Gardens Tampa Bay in Tampa, Florida, both Disney's Animal Kingdom and Gatorland in Orlando, Florida, Flamingo Land in North Yorkshire, England, and Six Flags Discovery Kingdom in Vallejo, California.

Sources of Animals

By the year 2000 most animals being displayed in zoos were the offspring of other zoo animals. This trend, however was and still is somewhat species-specific. When animals are transferred between zoos, they usually spend time in quarantine, and are given time to acclimatize to their new enclosures which are often designed to mimic their natural environment. For example, some species of penguins may require refrigerated enclosures. Guidelines on necessary care for such animals is published in the International Zoo Yearbook.

Justification

Conservation and Research

The position of most modern zoos in Australasia, Asia, Europe, and North America, particularly those with scientific societies, is that they display wild animals primarily for the conservation of endangered species, as well as for research purposes and education, and secondarily for the entertainment of visitors, an argument disputed by critics. The Zoological Society of London states in its charter that its aim is "the advancement of Zoology and Animal Physiology and the introduction of new and curious subjects of the Animal Kingdom." It maintains two research institutes, the Nuffield Institute of Comparative Medicine and the Wellcome Institute of Comparative Physiology. In the U.S., the Penrose Research Laboratory of the Philadelphia Zoo focuses on the study of comparative pathology. The World Association of Zoos and Aquariums produced its first conservation strategy in 1993, and in November 2004, it adopted a new strategy that sets out the aims and mission of zoological gardens of the 21st century.

Conservation programs all over the world fight to protect species from going extinct, but the unfortunate reality is, most conservation programs are underfunded and underrepresented. Conservation programs can struggle to fight bigger issues like habitat loss and illness. To rebuild degrading habitats takes major funding and a massive amount of time, both of which are scarce in conservation efforts. The current state of conservation programs cannot rely solely on situ (onsite conservation) plans alone, ex-situ (off-site conservation) is often needed. Off-site conservation relies on zoos, national parks, or other care facilities to support the rehabilitation of the animals and their populations. Zoos benefit conservation by providing suitable habitats and care to endangered animals. When properly regulated, they present a safe, clean environment for the animals to diversely repopulate. A study on amphibian conservation and zoos addressed these problems by writing:

> "Whilst addressing in situ threats, particularly habitat loss, degradation and fragmentation, is of primary importance; for many amphibian species in situ conservation alone will not be enough, especially in light of current un-mitigatable threats that can impact populations very rapidly such as chytridiomycosis [an infectious fungal disease]. Ex-situ programmes can complement in situ activities in a number of ways including maintaining genetically and demographically viable populations while threats are either better understood or mitigated in the wild".

The breeding of endangered species is coordinated by cooperative breeding programmes containing international studbooks and coordinators, who evaluate the roles of individual animals and institutions from a global or regional perspective, and there are regional programmes all over the world for the conservation of endangered species. In Africa, conservation is handled by the African Preservation Program (APP); in the U.S. and Canada by Species Survival Plans; in Australasia, by the Australasian Species Management Program; in Europe, by the European Endangered Species Program; and in Japan, South Asia, and South East Asia, by the Japanese Association of Zoos and Aquariums, the South Asian Zoo Association for Regional Cooperation, and the South East Asian Zoo Association.

Besides conservation of captive species, large zoos may form a suitable environment for wild native animals such as herons to live in or visit. A colony of black-crowned night herons has regularly summered at the National Zoo in Washington, D.C. for more than a century. Some zoos may provide information to visitors on wild animals visiting or living in the zoo, or encourage them by directing them to specific feeding or breeding platforms.

Roadside Zoos

In modern, well-regulated zoos, breeding is controlled to maintain a self-sustaining, global captive population. This is not the case in some less well-regulated zoos, often based in poorer regions. Overall "stock turnover" of animals during a year in a select group of poor zoos was reported as 20%-25% with 75% of wild caught apes dying in captivity within the first 20 months. The authors of the report stated that before successful breeding programs, the high mortality rate was the reason for the "massive scale of importations."

One 2-year study indicated that of 19,361 mammals that left accredited zoos in the U.S. between 1992 and 1998, 7,420 (38%) went to dealers, auctions, hunting ranches, unaccredited zoos and individuals, and game farms.

Animal Welfare Concerns

The welfare of zoo animals varies widely. Many zoos work to improve their animal enclosures and make it fit the animals' needs, although constraints such as size and expense make it difficult to create ideal captive environments for many species.

A study examining data collected over four decades found that polar bears, lions, tigers and cheetahs show evidence of stress in captivity. Zoos can be internment camps for animals, but also a place of refuge. A zoo can be considered an internment camp due to the insufficient enclosures that the animals have to live in. When an elephant is placed in a pen that is flat, has no tree, no other elephants and only a few plastic toys to play with; it can lead to boredom and foot problems. Also, animals can have a shorter life span when they are in these types of enclosures. Causes can be human diseases, materials in the cages, and possible escape attempts. However, when zoos take time to think about the animal's welfare, zoos can become a place of refuge. There are animals that are injured in the wild and are unable to survive on their own, but in the zoos they can live out the rest of their lives healthy and happy (McGaffin). In recent years, some zoos have chosen to stop showing their larger animals because they are simply unable to provide an adequate enclosure for them.

Moral Concerns

Some critics and many animal rights activists argue that zoo animals are treated as voyeuristic objects, rather than living creatures, and often suffer due to the transition from being free and wild to captivity. In the last 2 decades, European and North-American zoos, strongly depend on breeding within zoos, while decreasing the number of wild caught animals.

Behavioural Restriction

Many modern zoos attempt to improve animal welfare by providing more space and behavioural enrichments. This often involves housing the animals in naturalistic enclosures that allow the animals to express some of their natural behaviours, such as roaming and foraging. However, many animals remain in barren concrete enclosures or other minimally enriched cages.

Animals which naturally range over many km each day, or make seasonal migrations, are unable to perform these behaviors in zoo enclosures. For example, elephants usually travel approximately 45 km (28 mi) each day.

Abnormal Behaviour

Animals in zoos often exhibit behaviors that are abnormal in their frequency, intensity, or would not normally be part of their behavioural repertoire. These are usually indicative of stress. For example, elephants sometimes perform head-bobbing, bears sometimes pace repeatedly around the limits of their enclosure, wild cats sometimes groom themselves obsessively, and birds pluck out their own feathers. Some critics of zoos claim that the animals are always under physical and mental stress, regardless of the quality of care towards the animals. Elephants have been recorded displaying stereotypical behaviours in the form of swaying back and forth, trunk swaying or route tracing. This has been observed in 54% of individuals in UK zoos.

Shortened Longevity

Elephants in Japanese zoos have shorter lifespans than their wild counterparts at only 17 years, although other studies suggest that zoo elephants live as long as those in the wild. On the other hand, many other animals, such as reptiles, can live much longer than they would in the wild.

Climate Concerns

Climatic conditions can make it difficult to keep some animals in zoos in some locations. For example, Alaska Zoo had an elephant named Maggie. She was housed in a small, indoor enclosure because the outdoor temperature was too low.

Surplus Animals

Especially in large animals, a limited number of spaces are available in zoos. As a consequence, various management tools are used to preserve the space for the most "valuable" individuals and reduce the risk of inbreeding. Management of animal populations is typically through international organizations such as AZA and EAZA. Zoos have several different ways of managing the animal populations, such as moves between zoos, contraception, sale of excess animals and euthanization (culling).

Contraception can be effective, but may also have health repercussions and can be difficult (or even impossible) to reverse in some animals. Additionally, some species may lose their reproductive capability entirely if prevented from breeding for a period (whether through contraceptives or isolation), but further study is needed on the subject. Sale of surplus animals from zoos was once common and in some cases animals have ended up in substandard facilities. In recent decades the practice of selling animals from certified zoos has declined. A large number of animals are culled each year in zoos, but this is controversial. A highly publicized culling as part of population management was that of a healthy giraffe at Copenhagen Zoo in 2014. The zoo argued that its genes already were well-represented in captivity, making the giraffe unsuitable for future breeding. There were offers to adopt it and an online petition to save it had many thousand signatories, but the culling proceeded. Although zoos in some countries have been open about culling, the controversy of the subject and pressure from the public has resulted in others being closed. This stands in contrast to most zoos publicly announcing animal births. Furthermore, while many zoos are willing to cull smaller and/or low-profile animals, fewer are willing to do it with larger high-profile species.

Live Feeding and Baiting

In many countries, feeding live vertebrates to zoo animals is illegal, except in exceptional circumstances. For example, some snakes refuse to eat dead prey. However, in the Badaltearing Safari Park in China, visitors can throw live goats into the lion enclosure and watch them being eaten, or can purchase live chickens tied to bamboo rods for the equivalent of 2 dollars/euros to dangle into lion pens. Visitors can drive through the lion compound in buses with specially designed chutes which they can use to push live chickens into the enclosure. In the Xiongsen Bear and Tiger Mountain Village near Guilin in south-east China, live cows and pigs are thrown to tigers to amuse visitors.

In Qingdao zoo (Eastern China), visitors can engage in "tortoise baiting", where tortoises are kept inside small rooms with elastic bands around their necks so that they are unable to retract their heads. Visitors are allowed to throw coins at them. The marketing claim is that if a person hits one of the tortoises on the head and makes a wish, it will be fulfilled.

Regulation

In the United States, any public animal exhibit must be licensed and inspected by the Department of Agriculture, the Environmental Protection Agency, and the Occupational Safety and Health Administration. Depending on the animals they exhibit, the activities of zoos are regulated by laws including the Endangered Species Act, the Animal Welfare Act, the Migratory Bird Treaty Act of 1918 and others. Additionally, zoos in North America may choose to pursue accreditation by the Association of Zoos and Aquariums (AZA). To achieve accreditation, a zoo must pass an application and inspection process and meet or exceed the AZA's standards for animal health and welfare, fundraising, zoo staffing, and involvement in global conservation efforts. Inspection is performed by three experts (typically one veterinarian, one expert in animal care, and one expert in zoo management and operations) and then reviewed by a panel of twelve experts before accreditation is awarded. This accreditation process is repeated once every five years. The AZA estimates that there are approximately 2,400 animal exhibits operating under USDA license as of February 2007; fewer than 10% are accredited.

Europe

In April 1999, the European Union introduced a directive to strengthen the conservation role of zoos, making it a statutory requirement that they participate in conservation and education, and requiring all member states to set up systems for their licensing and inspection. Zoos are regulated in the UK by the Zoo Licensing Act of 1981, which came into effect in 1984. A zoo is defined as any "establishment where wild animals are kept for exhibition to which members of the public have access, with or without charge for admission, seven or more days in any period of twelve consecutive months", excluding circuses and pet shops. The Act requires that all zoos be inspected and licensed, and that animals kept in enclosures are provided with a suitable environment in which they can express most normal behavior.

Botanical Gardens

A botanical garden or botanic garden is a garden dedicated to the collection, cultivation, preservation and display a wide range of plants labelled with their botanical names. It may contain specialist plant collections such as cacti and other succulent plants, herb gardens, plants from particular parts of the world, and so on; there may be greenhouses, shadehouses, again with special collections such as tropical plants, alpine plants, or other exotic plants. Visitor services at a botanical garden might include tours, educational displays, art exhibitions, book rooms, open-air theatrical and musical performances, and other entertainment.

Botanical gardens are often run by universities or other scientific research organizations, and often have associated herbaria and research programmes in plant taxonomy or some other aspect of botanical science. In principle, their role is to maintain documented collections of living plants for the purposes of scientific research, conservation, display, and education, although this will depend on the resources available and the special interests pursued at each particular garden.

The origin of modern botanical gardens is generally traced to the appointment of professors of botany to the medical faculties of universities in 16th century Renaissance Italy, which also entailed the curation of a medicinal garden. However, the objectives, content, and audience of today's botanic gardens more closely resembles that of the grandiose gardens of antiquity and the educational garden of Theophrastus in the Lyceum of ancient Athens.

The early concern with medicinal plants changed in the 17th century to an interest in the new plant imports from explorations outside Europe as botany gradually established its independence from medicine. In the 18th century, systems of nomenclature and classification were devised by botanists working in the herbaria and universities associated with the gardens, these systems often being displayed in the gardens as educational "order beds". With the rapid rise of European imperialism in the late 18th century, botanic gardens were established in the tropics, and economic botany became a focus with the hub at the Royal Botanic Gardens, Kew, near London.

Over the years, botanical gardens, as cultural and scientific organisations, have responded to the interests of botany and horticulture. Nowadays, most botanical gardens display a mix of the themes mentioned and more; having a strong connection with the general public, there is the opportunity to provide visitors with information relating to the environmental issues being faced at the start of the 21st century, especially those relating to plant conservation and sustainability.

Many of the functions of botanical gardens have already been discussed which emphasise the scientific underpinning of botanical gardens with their focus on research, education and conservation. However, as multifaceted organisations, all sites have their own special interests. In a remarkable paper on the role of botanical gardens, Ferdinand Mueller (1825–1896), the director of the Royal Botanic Gardens, Melbourne (1852–1873), stated, "in all cases the objects [of a botanical garden] must be mainly scientific and predominantly instructive". He then detailed many of the objectives being pursued by the world's botanical gardens in the middle of the 19th century, when European gardens were at their height. Many of these are listed below to give a sense of the scope of botanical gardens' activities at that time, and the ways in which they differed from parks or what he called "public pleasure gardens":

- Availability of plants for scientific research.

- Display of plant diversity in form and use.

- Display of plants of particular regions (including local).

- Plants sometimes grown within their particular families.

- Plants grown for their seed or rarity.

- Major timber (American English: lumber) trees.

- Plants of economic significance.

- Glasshouse plants of different climates.

- All plants accurately labelled.

- Records kept of plants and their performance.

- Catalogues of holdings published periodically.

- Research facilities utilising the living collections.

- Studies in plant taxonomy.

- Examples of different vegetation types.

- Student education.

- A herbarium.

- Selection and introduction of ornamental and other plants to commerce.

- Studies of plant chemistry (phytochemistry).

- Report on the effects of plants on livestock.

- At least one collector maintained doing field work.

Botanical gardens must find a compromise between the need for peace and seclusion, while at the same time satisfying the public need for information and visitor services that include restaurants, information centres and sales areas that bring with them rubbish, noise, and hyperactivity. Attractive landscaping and planting design sometimes compete with scientific interests — with science now often taking second place. Some gardens are now heritage landscapes that are subject to constant demand for new exhibits and exemplary environmental management.

Many gardens now have plant shops selling flowers, herbs, and vegetable seedlings suitable for transplanting; many, like the UBC Botanical Garden and Centre for Plant Research and the Chicago Botanic Garden, have plant-breeding programs and introduce new plants to the horticultural trade.

Botanical gardens are still being built, such as the first botanical garden in Oman, which will be one of the largest gardens in the world. Once completed, it will house the first large-scale cloud forest in a huge glasshouse. Development of botanical gardens in China over recent years has been a remarkable, including the Hainan Botanical Garden of Tropical Economic Plants South China Botanical Garden at Guangzhou, the Xishuangbanna Botanical Garden of Tropical Plants and the Xiamen Botanic Garden, but in developed countries, many have closed for lack of financial support, this being especially true of botanical gardens attached to universities.

Botanical gardens have always responded to the interests and values of the day. If a single function were to be chosen from the early literature on botanical gardens, it would be their scientific endeavour and, flowing from this, their instructional value. In their formative years, botanical gardens were gardens for physicians and botanists, but then they progressively became more associated with ornamental horticulture and the needs of the general public. The scientific reputation of a botanical garden is now judged by the publications coming out of herbaria and similar facilities, not by its living collections. The interest in economic plants now has less relevance, and the concern with plant classification systems has all but disappeared, while a fascination with the curious, beautiful and new seems unlikely to diminish.

In recent times, the focus has been on creating an awareness of the threat to the Earth's ecosystems from human overpopulation and its consequent need for biological and physical resources. Botanical gardens provide an excellent medium for communication between the world of botanical science and the general public. Education programs can help the public develop greater environmental awareness by understanding the meaning and importance of ideas like conservation and sustainability.

ADVANTAGES OF BIODIVERSITY CONSERVATION

- Conservation of biological diversity leads to conservation of essential ecological diversity to preserve the continuity of food chains.

- The genetic diversity of plants and animals is preserved.

- It ensures the sustainable utilisation of life support systems on earth.

- It provides a vast knowledge of potential use to the scientific community.

- A reservoir of wild animals and plants is preserved, thus enabling them to be introduced, if need be, in the surrounding areas.

- Biological diversity provides immediate benefits to the society such as recreation and tourism.

- Biodiversity conservation serves as an insurance policy for the future.

Man and Biodiversity

Due to the growth in the human population, in production and consumption, over the last two centuries the natural ecosystems of our planet have been subjected to an impressive depletion of their biodiversity, with an overall decrease, measured by the Living Planet Index, equal to 30% from 1970 to 2005. Human activities have increased the rate of natural extinction and it is estimated that the current climate change will worsen the situation further. Biodiversity is important as a value itself, also because it contributes to human wellbeing: the vegetable components and the fauna in the forests are an important source of food for many local populations, they are a source of active ingredients (25% of the drugs), they contribute to increasing the revenue and freedom of choice of the local populations, they are remarkably important in social relations and conservation of the cultural heritage.

References

- Why-is-biodiversity-important-who-cares: globalissues.Org, retrieved 17 march, 2019

- Cardinale, bradley. J.; Et al. (March 2011). "The functional role of producer diversity in ecosystems". American journal of botany. 98 (3): 572–592. Doi:10.3732/Ajb.1000364. Pmid 21613148.

- Conservation-of-biodiversity, biodiversity-1, environment, energy: vikaspedia.In, retrieved 25 august, 2019

- Letourneau, deborah k. (1 January 2011). "Does plant diversity benefit agroecosystems? A synthetic review". Ecological applications. 21 (1): 9–21. Doi:10.1890/09-2026.1.

- "History of the national parks". Association of national park authorities. Archived from the original on 21 april 2013. Retrieved 12 november 2012

- Bonnett, a. (2016). The geography of nostalgia: global and local perspectives on modernity and loss. Routledge. P. 68. Isbn 978-1-315-88297-0

- Conservation-of-biodiversity, biodiversity-1, environment, energy: vikaspedia.In, retrieved 13 february, 2019

- "Mission statement". Bgci.Org. Botanic gardens conservation international. Archived from the original on 28 october 2011. Retrieved 8 november 2011.

- Man-and-biodiversity, the-value-of-biodiversity1, biodiversity1, argomento: eniscuola.Net, retrieved 4 february, 2019

- Braverman, irus (2012). Zooland: the institution of captivity, stanford university press, stanford. Isbn 9780804783576

Five Kingdom Classification

The organisms have been classified into five kingdoms on the basis of likeness and variances among them. Monera, protist, fungus, plant and animal are the five kingdoms of organisms. The topics elaborated in this chapter will help in gaining a better perspective about these classifications of organisms.

The five kingdom classification that we see today was not the initial result of the classification of living organisms. Carolus Linnaeus first came up with a two kingdom classification which included only kingdom Plantae and kingdom Animalia.

The two kingdom classification lasted for a very long time but did not last forever because it did not take into account many major parameters while classifying. There was no differentiation of the eukaryotes and prokaryotes; neither unicellular and multicellular; nor photosynthetic and the non-photosynthetic.

Putting all the organisms in either plant or animal kingdom was insufficient because there were a lot of organisms which could not be classified as either plants or animals.

All these confusions led to a new mode of classification which had to take into account cell structure, the presence of cell wall, mode of reproduction and mode of nutrition. As a result, R H Whittaker came up with the concept of five kingdom classification.

The five kingdom classification of living organisms included the following kingdoms:

Kingdom Monera

The bacteria are categorized underneath the Kingdom Monera.

Features of Monerans

They possess the following important features:

- Bacteria occur everywhere and they are microscopic in nature.
- They possess a cell wall and are prokaryotic.
- The cell wall is formed of amino acids and polysaccharides.
- Bacteria can be heterotrophic and autotrophic.
- The heterotrophic bacteria can be parasitic or saprophytic. The autotrophic bacteria can be chemosynthetic or photosynthetic.

Types of Monerans

Bacteria can be classified into four types based on their shape:

- Coccus (pl.: cocci)- These bacteria are spherical in shape.

- Bacillus (pl.: bacilli) – These bacteria are rod-shaped.

- Vibrium (pl.: vibrio) – These bacteria are comma-shaped bacteria.

- Spirillum (pl.: spirilla)- These bacteria are spiral-shaped bacteria.

Monera has since been divided into Archaebacteria and Eubacteria.

Kingdom Protista

Features of Protista

Protista has the following important features:

- They are unicellular and eukaryotic organisms.

- Some of them have cilia or flagella for mobility.

- Sexual reproduction is by a process of cell fusion and zygote formation.

Sub-groups of Protista

Kingdom Protista is categorized into subsequent groups:

- Chrysophytes: The golden algae (desmids) and diatoms fall under this group. They are found in marine and freshwater habitats.

- Dinoflagellates: They are usually photosynthetic and marine. The colour they appear is dependent on the key pigments in their cells; they appear red, blue, brown, green or yellow.

- Euglenoids: Most of them live in freshwater habitation in motionless water. The cell wall is absent in them, instead, there is a protein-rich layer called pellicle.

- Slime Moulds: These are saprophytic. The body moves along putrefying leaves and twigs and nourishes itself on organic material. Under favourable surroundings, they form an accumulation and were called Plasmodial slime moulds.

- Protozoans: They are heterotrophs and survive either as parasites or predators.

Kingdom Fungi

The kingdom fungi include moulds, mushroom, yeast etc. They show a variety of applications in domestic as well as commercial purposes.

Features of Kingdom Fungi:

- The fungi are filamentous, excluding yeast (single-celled).

- Their figure comprises of slender, long thread-like constructions called hyphae. The web of hyphae is called mycelium.

- Some of the hyphae are unbroken tubes which are jam-packed with multinucleated cytoplasm. Such hyphae are labelled Coenocytic hyphae.

- The other type of hyphae has cross-walls or septae.

- The cell wall of fungi is composed of polysaccharides and chitin.

- Most of the fungi are saprophytes and are heterotrophic.

- Some of the fungi also survive as symbionts. Some are parasites. Some of the symbiont fungi live in association with algae, like lichens. Some of the symbiont fungi live in association with roots of higher plants, as mycorrhiza.

Kingdom Plantae

Features of Kingdom Plantae:

- The kingdom Plantae is filled with all eukaryotes which have chloroplast.

- Most of them are autotrophic in nature, but some are heterotrophic as well.

- The Cell wall mainly comprises of cellulose.

- Plants have two distinct phases in their lifecycle. These phases alternate with each other. The diploid saprophytic and the haploid gametophytic phase. The lengths of the diploid and haploid phases vary among dissimilar groups of plants. Alternation of Generation is what this phenomenon is called.

Kingdom Animalia

Features of Kingdom Animalia:

- All multicellular eukaryotes which are heterotrophs and lack cell wall are set aside under this kingdom.

- The animals are directly or indirectly dependent on food on plants. Their mode of nutrition is holozoic. Holozoic nutrition encompasses ingestion of food and then the use of an internal cavity for digestion of food.

- Many of the animals are adept for locomotion.

- They reproduce by sexual mode of reproduction.

The five kingdom classification of living organisms took a lot into consideration and is till now the most efficient system.

The older system of classification was based only on one single characteristic according to which two highly varied organisms were grouped together. For example, the fungi and plants were placed in the same group based on the presence of the cell wall. In the same way, unicellular and multicellular organisms were also grouped together.

Therefore, all the organisms were classified again into the five kingdoms known as the five kingdom classification, starting with Monera, where all the prokaryotic unicellular organisms were placed together.

Following that, all the eukaryotic unicellular organisms were placed under the kingdom Protista.

The organisms were then classified based on the presence and absence of a cell wall. The ones' without the cell wall were classified under kingdom Animalia and the ones' with cell wall were classified under kingdom Plantae.

The organisms under kingdom Plantae were further classified into photosynthetic and nonphotosynthetic which included Plantae and fungi respectively.

This system of classification of living organisms is better than following the older classification of plants and animals because it eradicated the confusion of putting one species in two different kingdoms.

In biology, kingdom is the second highest taxonomic rank, just below domain. Kingdoms are divided into smaller groups called phyla.

Traditionally, some textbooks from the United States used a system of six kingdoms (Animalia, Plantae, Fungi, Protista, Archaea/Archaebacteria, and Bacteria/Eubacteria) while textbooks in countries like Great Britain, India, Greece, Australia, Latin America and other countries used five kingdoms (Animalia, Plantae, Fungi, Protista and Monera).

Some recent classifications based on modern cladistics have explicitly abandoned the term "kingdom", noting that the traditional kingdoms are not monophyletic, i.e., do not consist of all the descendants of a common ancestor.

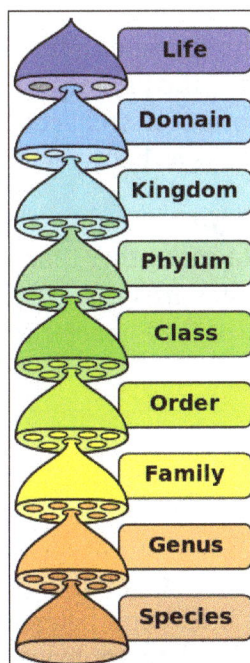

The hierarchy of biological classification's eight major
taxonomic ranks. A domain contains one or more kingdoms.

Modern View

Three Domains of Life

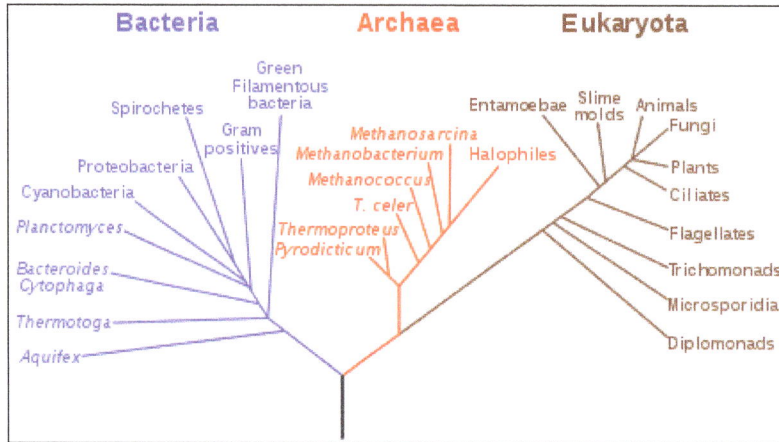

A phylogenetic tree based on rRNA data showing Woese's three-domain system. All smaller branches can be considered kingdoms.

From around the mid-1970s onwards, there was an increasing emphasis on comparisons of genes at the molecular level (initially ribosomal RNA genes) as the primary factor in classification; genetic similarity was stressed over outward appearances and behavior. Taxonomic ranks, including kingdoms, were to be groups of organisms with a common ancestor, whether monophyletic (all descendants of a common ancestor) or paraphyletic (only some descendants of a common ancestor). Based on such RNA studies, Carl Woese thought life could be divided into three large divisions and referred to them as the "three primary kingdom" model or "urkingdom" model. In 1990, the name "domain" was proposed for the highest rank. This term represents a synonym for the category of dominion (lat. dominium), introduced by Moore in 1974. Unlike Moore, Woese et al. did not suggest a Latin term for this category, which represents a further argument supporting the accurately introduced term dominion. Woese divided the prokaryotes (previously classified as the Kingdom Monera) into two groups, called Eubacteria and Archaebacteria, stressing that there was as much genetic difference between these two groups as between either of them and all eukaryotes.

According to genetic data, although eukaryote groups such as plants, fungi, and animals may look different, they are more closely related to each other than they are to either the Eubacteria or Archaea. It was also found that the eukaryotes are more closely related to the Archaea than they are to the Eubacteria. Although the primacy of the Eubacteria-Archaea divide has been questioned, it has been upheld by subsequent research.

Kingdoms of the Eukaryota

Simpson and Roger noted that the Protista were "a grab-bag for all eukaryotes that are not animals, plants or fungi". They held that only monophyletic groups should be accepted as formal ranks in a classification and that – while this approach had been impractical previously (necessitating "literally dozens of eukaryotic 'kingdoms'") – it had now become possible to divide the eukaryotes into "just a few major groups that are probably all monophyletic".

On this basis, the diagram above showed the real "kingdoms" (their quotation marks) of the eukaryotes.

A classification which followed this approach was produced in 2005 for the International Society of Protistologists, by a committee which "worked in collaboration with specialists from many societies". It divided the eukaryotes into the same six "supergroups". The published classification deliberately did not use formal taxonomic ranks, including that of "kingdom".

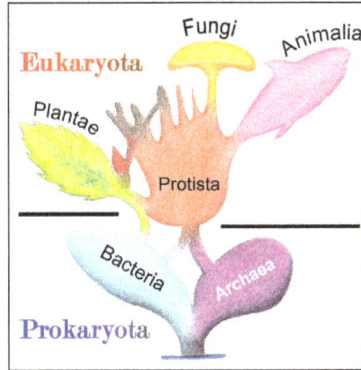

Phylogenetic and symbiogenetic tree of living organisms, showing the origins of eukaryotes & prokaryotes.

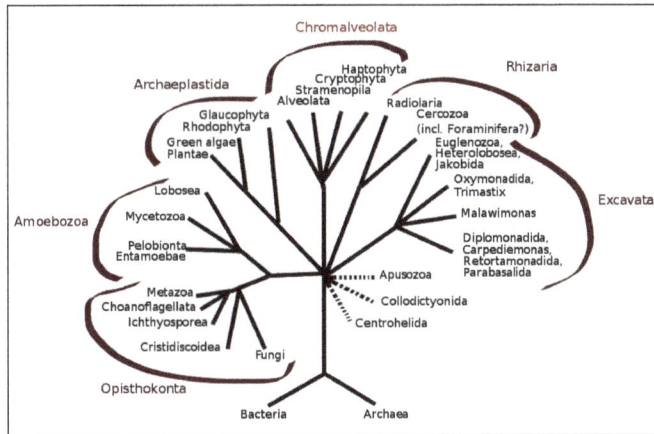

One hypothesis of eukaryotic relationships, modified from Simpson and Roger.

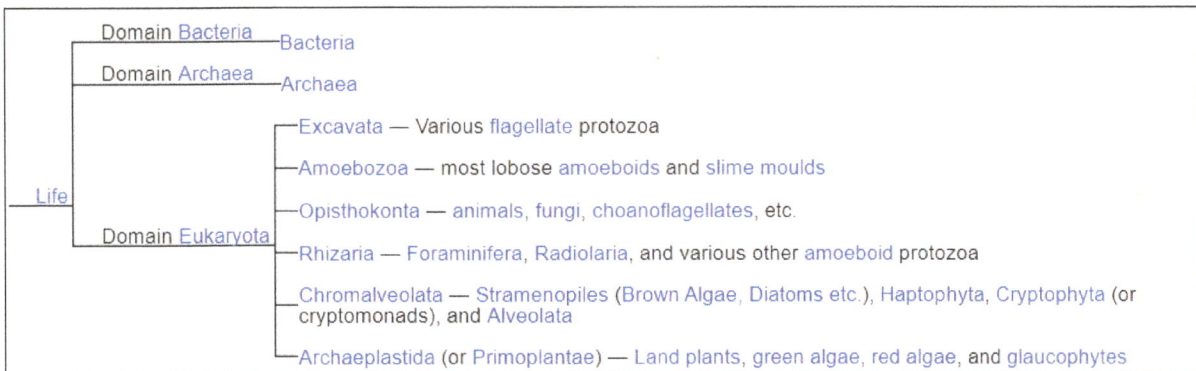

In this system the multicellular animals (Metazoa) are descended from the same ancestor as both the unicellular choanoflagellates and the fungi which form the Opisthokonta. Plants are thought to be more distantly related to animals and fungi.

However, in the same year as the International Society of Protistologists' classification was published, doubts were being expressed as to whether some of these supergroups were monophyletic, particularly the Chromalveolata, and a review in 2006 noted the lack of evidence for several of the six proposed supergroups.

As of 2010, there is widespread agreement that the Rhizaria belong with the Stramenopiles and the Alveolata, in a clade dubbed the SAR supergroup, so that Rhizaria is not one of the main eukaryote groups. Beyond this, there does not appear to be a consensus. Rogozin et al. in 2009 noted that "The deep phylogeny of eukaryotes is an extremely difficult and controversial problem." As of December 2010, there appears to be a consensus that the six supergroup model proposed in 2005 does not reflect the true phylogeny of the eukaryotes and hence how they should be classified, although there is no agreement as to the model which should replace it.

Two Life Forms: Animals and Plants

The classification of living things into animals and plants is an ancient one. Aristotle classified animal species in his History of Animals, while his pupil Theophrastus wrote a parallel work, the Historia Plantarum, on plants.

Three Kingdoms: Animal, Vegetable and Mineral

Carl Linnaeus laid the foundations for modern biological nomenclature, now regulated by the Nomenclature Codes, in 1735. He distinguished two kingdoms of living things: Regnum Animale ('animal kingdom') and Regnum Vegetabile ('vegetable kingdom', for plants). Linnaeus also included minerals in his classification system, placing them in a third kingdom, Regnum Lapideum.

Three Kingdoms: Plants, Protists, and Animals

In 1674, Antonie van Leeuwenhoek, often called the "father of microscopy", sent the Royal Society of London a copy of his first observations of microscopic single-celled organisms. Until then, the existence of such microscopic organisms was entirely unknown. Despite this, Linnaeus did not include any microscopic creatures in his original taxonomy.

At first, microscopic organisms were classified within the animal and plant kingdoms. However, by the mid-19th century, it had become clear to many that "the existing dichotomy of the plant and animal kingdoms [had become] rapidly blurred at its boundaries and outmoded". In 1866, Ernst Haeckel proposed a third kingdom of life, the Protista, for "neutral organisms" which were neither animal nor plant. Haeckel revised the content of this kingdom a number of times before settling on a division based on whether organisms were unicellular (Protista) or multicellular (animals and plants).

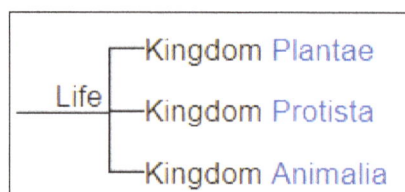

Four Kingdoms

The development of the electron microscope revealed important distinctions between those unicellular organisms whose cells do not have a distinct nucleus (prokaryotes) and those unicellular and multicellular organisms whose cells do have a distinct nucleus (eukaryotes). In 1938, Herbert F. Copeland proposed a four-kingdom classification, elevating the protist classes of bacteria (Monera) and blue-green algae (Phycochromacea) to phyla in the novel Kingdom Monera.

Domain, Empires and Four or Five Kingdoms

The importance of the distinction between prokaryotes and eukaryotes gradually became apparent. In the 1960s, Stanier and van Niel popularised Édouard Chatton's much earlier proposal to recognise this division in a formal classification. This required the creation, for the first time, of a rank above kingdom, a superkingdom or empire, later called a domain.

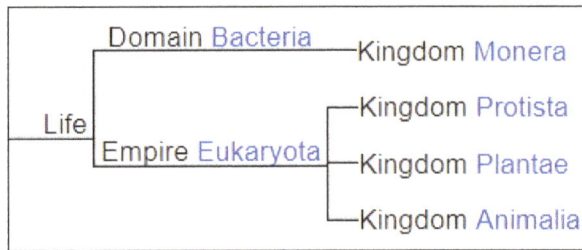

The differences between fungi and other organisms regarded as plants had long been recognised by some; Haeckel had moved the fungi out of Plantae into Protista after his original classification, but was largely ignored in this separation by scientists of his time. Robert Whittaker recognized an additional kingdom for the Fungi. The resulting five-kingdom system, proposed in 1969 by Whittaker, has become a popular standard and with some refinement is still used in many works and forms the basis for new multi-kingdom systems. It is based mainly upon differences in nutrition; his Plantae were mostly multicellular autotrophs, his Animalia multicellular heterotrophs, and his Fungi multicellular saprotrophs. The remaining two kingdoms, Protista and Monera, included unicellular and simple cellular colonies. The five kingdom system may be combined with the two empire system:

In the Whittaker system, Plantae included some algae. In other systems, such as Lynn Margulis's system of five kingdoms—animals, plants, bacteria (prokaryotes), fungi, and protoctists—the plants included just the land plants (Embryophyta).

Despite the development from two kingdoms to five among most scientists, some authors as late as 1975 continued to employ a traditional two-kingdom system of animals and plants, dividing the plant kingdom into Subkingdoms Prokaryota (bacteria and cyanophytes), Mycota (fungi and supposed relatives), and Chlorota (algae and land plants).

Eight Kingdoms

Thomas Cavalier-Smith thought at first, as was almost the consensus at that time, that the difference between eubacteria and archaebacteria was so great (particularly considering the genetic distance of ribosomal genes) that they needed to be separated into two different kingdoms, hence splitting the empire Bacteria into two kingdoms. He then divided Eubacteria into two subkingdoms: Negibacteria (Gram negative bacteria) and Posibacteria (Gram positive bacteria).

Technological advances in electron microscopy allowed the separation of the Chromista from the Plantae kingdom. Indeed, the chloroplast of the chromists is located in the lumen of the endoplasmic reticulum instead of in the cytosol. Moreover, only chromists contain chlorophyll c. Since then, many non-photosynthetic phyla of protists, thought to have secondarily lost their chloroplasts, were integrated into the kingdom Chromista.

Finally, some protists lacking mitochondria were discovered. As mitochondria were known to be the result of the endosymbiosis of a proteobacterium, it was thought that these amitochondriate eukaryotes were primitively so, marking an important step in eukaryogenesis. As a result, these amitochondriate protists were separated from the protist kingdom, giving rise to the, at the same time, superkingdom and kingdom Archezoa. This was known as the Archezoa hypothesis. This superkingdom was opposed to the Metakaryota superkingdom, grouping together the five other eukaryotic kingdoms (Animalia, Protozoa, Fungi, Plantae and Chromista).

Six Kingdoms

In 1998, Cavalier-Smith published a six-kingdom model, which has been revised in subsequent papers. The version published in 2009 is shown below. Cavalier-Smith no longer accepts the importance of the fundamental eubacteria–archaebacteria divide put forward by Woese and others and supported by recent research. His Kingdom Bacteria includes Archaebacteria as a phylum of the subkingdom Unibacteria which comprises only one other phylum: the Posibacteria. The two subkingdoms Unibacteria and Negibacteria of kingdom Bacteria (sole kingdom of empire Prokaryota) are distinguished according to their membrane topologies. The bimembranous-unimembranous transition is thought to be far more fundamental than the long branch of genetic distance of Archaebacteria, viewed as having no particular biological significance. Cavalier-Smith does not accept the requirement for taxa to be monophyletic ("holophyletic" in his terminology) to be valid. He defines Prokaryota, Bacteria, Negibacteria, Unibacteria, and Posibacteria as valid paraphyla (therefore "monophyletic" in the sense he uses this term) taxa, marking important innovations of biological significance (in regard of the concept of biological niche).

In the same way, his paraphyletic kingdom Protozoa includes the ancestors of Animalia, Fungi, Plantae, and Chromista. The advances of phylogenetic studies allowed Cavalier-Smith to realize that all the phyla thought to be archezoans (i.e. primitively amitochondriate eukaryotes) had in fact secondarily lost their mitochondria, typically by transforming them into new organelles: Hydrogenosomes. This means that all living eukaryotes are in fact metakaryotes, according to the significance of the term given by Cavalier-Smith. Some of the members of the defunct kingdom

Archezoa, like the phylum Microsporidia, were reclassified into kingdom Fungi. Others were re-classified in kingdom Protozoa like Metamonada which is now part of infrakingdom Excavata.

Because Cavalier-Smith allows paraphyly, the diagram below is an 'organization chart', not an 'ancestor chart', and does not represent evolutionary tree.

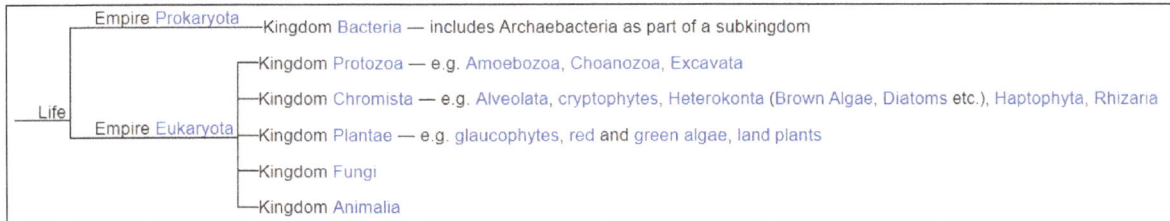

```
                Empire Prokaryota ──Kingdom Bacteria — includes Archaebacteria as part of a subkingdom
                              ┌──Kingdom Protozoa — e.g. Amoebozoa, Choanozoa, Excavata
                              ├──Kingdom Chromista — e.g. Alveolata, cryptophytes, Heterokonta (Brown Algae, Diatoms etc.), Haptophyta, Rhizaria
       Life │                ├──Kingdom Plantae — e.g. glaucophytes, red and green algae, land plants
                Empire Eukaryota──Kingdom Plantae
                              ├──Kingdom Fungi
                              └──Kingdom Animalia
```

Seven Kingdoms

The kingdom-level classification of life is still widely employed as a useful way of grouping organisms, notwithstanding some problems with this approach:

- Kingdoms such as Bacteria represent grades rather than clades, and so are rejected by phylogenetic classification systems.

- The most recent research does not support the classification of the eukaryotes into any of the standard systems. As of April 2010, no set of kingdoms is sufficiently supported by research to attain widespread acceptance. In 2009, Andrew Roger and Alastair Simpson emphasized the need for diligence in analyzing new discoveries: "With the current pace of change in our understanding of the eukaryote tree of life, we should proceed with caution."

Viruses

There is ongoing debate as to whether viruses can be included in the tree of life. The ten arguments against include the fact that they are obligate intracellular parasites that lack metabolism and are not capable of replication outside of a host cell. Another argument is that their placement in the tree would be problematic, since it is suspected that viruses have arisen multiple times, and they have a penchant for harvesting nucleotide sequences from their hosts.

On the other hand, arguments favor their inclusion. One comes from the discovery of unusually large and complex viruses, such as Mimivirus, that possess typical cellular genes.

CLASSIFICATION OF FIVE KINGDOM

Monera

Monera is a kingdom that contains unicellular organisms with a prokaryotic cell organization (having no nuclear membrane), such as bacteria. They are single-celled organisms with no true nuclear membrane (prokaryotic organisms).

The taxon Monera was first proposed as a phylum by Ernst Haeckel in 1866. Subsequently, the phylum was elevated to the rank of kingdom in 1925 by Édouard Chatton. The last commonly accepted mega-classification with the taxon Monera was the five-kingdom classification system established by Robert Whittaker in 1969.

Under the three-domain system of taxonomy, introduced by Carl Woese in 1977, which reflects the evolutionary history of life, the organisms found in kingdom Monera have been divided into two domains, Archaea and Bacteria (with Eukarya as the third domain). Furthermore, the taxon Monera is paraphyletic (does not include all descendants of their most-recent common ancestor), as Archaea and Eukarya are currently believed to be more closely related than either is to Bacteria. The term "moneran" is the informal name of members of this group and is still sometimes used (as is the term "prokaryote") to denote a member of either domain.

Most bacteria were classified under Monera; however, Cyanobacteria (often called the blue-green algae) were initially classified under Plantae due to their ability to photosynthesize.

Haeckel's Classification

Traditionally the natural world was classified as animal, vegetable, or mineral as in Systema Naturae. After the development of the microscope, attempts were made to fit microscopic organisms into either the plant or animal kingdoms. In 1675, Antonie van Leeuwenhoek discovered bacteria and called them "animalcules", assigning them to the class Vermes of the Animalia. Due to the limited tools — the sole references for this group were shape, behaviour, and habitat — the description of genera and their classification was extremely limited, which was accentuated by the perceived lack of importance of the group.

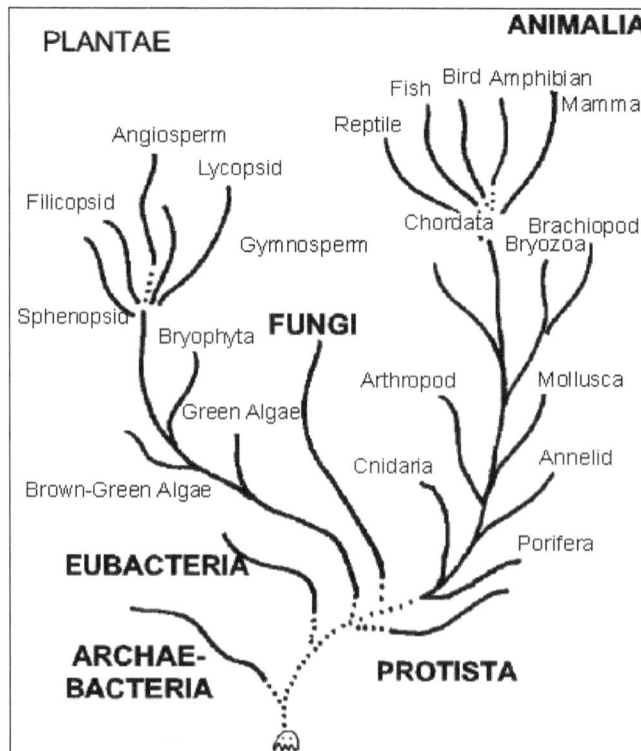

Tree of Life in Generelle Morphologie der Organismen.

Ten years after The Origin of Species by Charles Darwin, in 1866 Ernst Haeckel, a supporter of evolution, proposed a three-kingdom system that added the Protista as a new kingdom that contained most microscopic organisms. One of his eight major divisions of Protista was composed of the monerans (called Moneres by Haeckel), which he defined as completely structure-less and homogeneous organisms, consisting only of a piece of plasma. Haeckel's Monera included not only bacterial groups of early discovery but also several small eukaryotic organisms; in fact the genus Vibrio is the only bacterial genus explicitly assigned to the phylum, while others are mentioned indirectly, which led Copeland to speculate that Haeckel considered all bacteria to belong to the genus Vibrio, ignoring other bacterial genera. One notable exception were the members of the modern phylum Cyanobacteria, such as Nostoc, which were placed in the phylum Archephyta of Algae (vide infra: Blue-green algae).

The Neolatin noun Monera and the German noun Moneren/Moneres are derived from the ancient Greek noun moneres, which Haeckel stated meant "simple"; however, it actually means "single, solitary". Haeckel also describes the protist genus Monas in the two pages about Monera in his 1866 book. The informal name of a member of the Monera was initially moneron, but later moneran was used.

Due to its lack of features, the phylum was not fully subdivided, but the genera therein were divided into two groups:

- Die Gymnomoneren (no envelope): Gymnomonera.

  - Protogenes — such as Protogenes primordialis, an unidentified amoeba (eukaryote) and not a bacterium.

  - Protamaeba — an incorrectly described/fabricated species.

- Vibrio — a genus of comma-shaped bacteria first described in 1854.

- Bacterium — a genus of rod-shaped bacteria first described in 1828. Haeckel does not explicitly assign this genus to the Monera.

- Bacillus — a genus of spore-forming rod-shaped bacteria first described in 1835 Haeckel does not explicitly assign this genus to the Monera kingdom.

- Spirochaeta — thin spiral-shaped bacteria first described in 1835 Haeckel does not explicitly assign this genus to the Monera.

- Spirillum — spiral-shaped bacteria first described in 1832 Haeckel does not explicitly assign this genus to the Monera.

- Haeckel does provide a comprehensive list.

- Die Lepomoneren (with envelope): Lepomonera.

  - Protomonas — identified to a synonym of Monas, a flagellated protozoan, and not a bacterium. The name was reused in 1984 for an unrelated genus of bacteria.

  - Vampyrella — now classed as a eukaryote and not a bacterium.

Subsequent Classifications

Like Protista, the Monera classification was not fully followed at first and several different ranks were used and located with animals, plants, protists or fungi. Furthermore, Häkel's classification lacked specificity and was not exhaustive — it in fact covers only a few pages—, consequently a lot of confusion arose even to the point that the Monera did not contain bacterial genera and others according to Huxley. They were first recognized as a kingdom by Enderlein in 1925.

C. Von Nägeli who classified non-phototrophic Bacteria as the class Schizomycetes.

The class Schizomycetes was then emended by Walter Migula (along with the coinage of the genus Pseudomonas in 1894) and others. This term was in dominant use even in 1916 as reported by Robert Earle Buchanan, as it had priority over other terms such as Monera. However, starting with Ferdinand Cohn in 1872 the term bacteria (or in German Bacterien) became prominently used to informally describe this group of species without a nucleus: Bacterium was in fact a genus created in 1828 by Christian Gottfried Ehrenberg Additionally, Cohn divided the bacteria according to shape namely:

- Spherobacteria for the cocci.

- Microbacteria for the short, non-filamentous rods.

- Desmobacteria for the longer, filamentous rods and Spirobacteria for the spiral forms.

Successively, Cohn created the Schizophyta of Plants, which contained the non-photrophic bacteria in the family Schizomycetes and the phototrophic bacteria (blue green algae/Cyanobacteria) in the Schizophyceae. This union of blue green algae and Bacteria was much later followed by Haeckel, who classified the two families in a revised phylum Monera in the Protista.

Rise to Prominence

The term Monera became well established in the 20s and 30s when to rightfully increase the importance of the difference between species with a nucleus and without, In 1925, Édouard Chatton divided all living organisms into two empires Prokaryotes and Eukaryotes: the Kingdom Monera being the sole member of the Prokaryotes empire.

The anthropic importance of the crown group of animals, plants and fungi was hard to depose; consequently, several other megaclassification schemes ignored on the empire rank but maintained the kingdom Monera consisting of bacteria, such Copeland in 1938 and Whittaker in 1969. The latter classification system was widely followed, in which Robert Whittaker proposed a five kingdom system for classification of living organisms. Whittaker's system placed most single celled organisms into either the prokaryotic Monera or the eukaryotic Protista. The other three kingdoms in his system were the eukaryotic Fungi, Animalia, and Plantae. Whittaker, however, did not believe that all his kingdoms were monophyletic. Whittaker subdivided the kingdom into two branches containing several phyla:

- Myxomonera branch

  ° Cyanophyta, now called Cyanobacteria.

  ° Myxobacteria.

- Mastigomonera branch

  ◦ Eubacteriae.

  ◦ Actinomycota.

  ◦ Spirochaetae.

Alternative commonly followed subdivision systems were based on Gram stains. This culminated in the Gibbons and Murray classification of 1978:

Gracilicutes (Gram Negative)

- Photobacteria (photosynthetic): class Oxyphotobacteriae (water as electron acceptor, includes the order Cyanobacteriales = blue green algae, now phylum Cyanobacteria) and class Anoxyphotobacteriae (anaerobic phototrophs, orders: Rhodospirillales and Chlorobiales.

- Scotobacteria (non-photosynthetic, now the Proteobacteria and other gram negative non-photosynthetic phyla).

Firmacutes (Gram Positive, Subsequently Corrected to Firmicutes)

- Several orders such as Bacillales and Actinomycetales (now in the phylum Actinobacteria).

- Mollicutes (gram variable, e.g. Mycoplasma).

- Mendocutes (uneven gram stain, "methanogenic bacteria" now known as the Archaea).

Three-Domain System

In 1977, a PNAS paper by Carl Woese and George Fox demonstrated that the archaea (initially called archaebacteria) are not significantly closer in relationship to the bacteria than they are to eukaryotes. The paper received front-page coverage in The New York Times, and great controversy initially. The conclusions have since become accepted, leading to replacement of the kingdom Monera with the two domains Bacteria and Archaea. A minority of scientists, including Thomas Cavalier-Smith, continue to reject the widely accepted division between these two groups. Cavalier-Smith has published classifications in which the archaebacteria are part of a subkingdom of the Kingdom Bacteria.

Blue-Green Algae

Although it was generally accepted that one could distinguish prokaryotes from eukaryotes on the basis of the presence of a nucleus, mitosis versus binary fission as a way of reproducing, size, and other traits, the monophyly of the kingdom Monera (or for that matter, whether classification should be according to phylogeny) was controversial for many decades. Although distinguishing between prokaryotes from eukaryotes as a fundamental distinction is often credited to a 1937 paper by Édouard Chatton (little noted until 1962), he did not emphasize this distinction more than other biologists of his era. Roger Stanier and C. B. van Niel believed that the bacteria (a term which at the time did not include blue-green algae) and the blue-green algae had a single origin, a conviction that culminated in Stanier writing in a letter in 1970, "I think it is now quite evident that the blue-green algae are not distinguishable from bacteria by any fundamental feature of their cellular organization". Other researchers, such as E. G. Pringsheim writing in 1949, suspected separate origins for bacteria and blue-green algae. In 1974, the influential Bergey's Manual published a new edition coining the term cyanobacteria to refer to what had been called blue-green algae, marking the acceptance of this group within the Monera.

Protist

A protist is any eukaryotic organism (one with cells containing a nucleus) that is not an animal, plant or fungus. The protists do not form a natural group, or clade, since they exclude certain eukaryotes; but, like algae or invertebrates, they are often grouped together for convenience. In some systems of biological classification, such as the popular five-kingdom scheme proposed by Robert Whittaker in 1969, the protists make up a kingdom called Protista, composed of "organisms which are unicellular or unicellular-colonial and which form no tissues".

Besides their relatively simple levels of organization, protists do not necessarily have much in common. When used, the term "protists" is now considered to mean a paraphyletic assemblage of similar-appearing but diverse taxa (biological groups); these taxa do not have an exclusive common ancestor beyond being composed of eukaryotes and have different life cycles, trophic levels, modes of locomotion and cellular structures. In the classification system of Lynn Margulis, the term protist is reserved for microscopic organisms, while the more inclusive term Protoctista is applied to a biological kingdom that includes certain large multicellular eukaryotes, such as kelp, red algae and slime molds. Others use the term protist more broadly, to encompass both microbial eukaryotes and macroscopic organisms that do not fit into the other traditional kingdoms.

In cladistic systems (classifications based on common ancestry), there are no equivalents to the taxa Protista or Protoctista, both terms referring to a paraphyletic group that spans the entire eukaryotic tree of life. In cladistic classification, the contents of Protista are distributed among various supergroups (SAR, such as protozoa and some algae, Archaeplastida, such as land plants and some algae, Excavata, which are a group of unicellular organisms, and Opisthokonta, such as animals and fungi, etc.). "Protista", "Protoctista" and "Protozoa" are considered obsolete. However, the term "protist" continues to be used informally as a catch-all term for unicellular eukaryotic microorganisms. For example, the word "protist pathogen" may be used to denote any disease-causing microbe that is not bacteria, virus, viroid, prion, or metazoa.

Subdivisions

The term protista was first used by Ernst Haeckel in 1866. Protists were traditionally subdivided into several groups based on similarities to the "higher" kingdoms such as:

Protozoa

These unicellular "animal-like" (heterotrophic, and sometimes parasitic) organisms are further sub-divided based on characteristics such as motility, such as the (flagellated) Flagellata, the (ciliated) Ciliophora, the (phagocytic) amoeba, and the (spore-forming) Sporozoa.

Protophyta

These "plant-like" (autotrophic) organisms are composed mostly of unicellular algae.

Molds

Slime molds and water molds are "fungus-like" (saprophytic) organisms.

Some protists, sometimes called ambiregnal protists, have been considered to be both protozoa and algae or fungi (e.g., slime molds and flagellated algae), and names for these have been published under either or both of the ICN and the ICZN. Conflicts, such as these – for example the dual-classification of Euglenids and Dinobryons, which are mixotrophic – is an example of why the kingdom Protista was adopted.

These traditional subdivisions, largely based on superficial commonalities, have been replaced by classifications based on phylogenetics (evolutionary relatedness among organisms). Molecular analyses in modern taxonomy have been used to redistribute former members of this group into diverse and sometimes distantly related phyla. For instance, the water molds are now considered to be closely related to photosynthetic organisms such as Brown algae and Diatoms, the slime molds are grouped mainly under Amoebozoa, and the Amoebozoa itself includes only a subset of "Amoeba" group, and significant number of erstwhile "Amoeboid" genera are distributed among Rhizarians and other Phyla.

However, the older terms are still used as informal names to describe the morphology and ecology of various protists. For example, the term protozoa is used to refer to heterotrophic species of protists that do not form filaments.

Classification

Among the pioneers in the study of the protists, which were almost ignored by Linnaeus except for some genera (e.g., Vorticella, Chaos, Volvox, Corallina, Conferva, Ulva, Chara, Fucus) were Leeuwenhoek, O. F. Müller, C. G. Ehrenberg and Félix Dujardin. The first groups used to classify microscopic organism were the Animalcules and the Infusoria. In 1818, the German naturalist Georg August Goldfuss introduced the word Protozoa to refer to organisms such as ciliates and corals. After the cell theory of Schwann and Schleiden (1838–39), this group was modified in 1848 by Carl von Siebold to include only animal-like unicellular organisms, such as foraminifera and amoebae. The formal taxonomic category Protoctista was first proposed in the early 1860s by John Hogg, who argued that the protists should include what he saw as primitive unicellular forms of both plants and animals. He defined the Protoctista as a "fourth kingdom of nature", in addition to the then-traditional kingdoms of plants, animals and minerals. The kingdom of minerals was later removed from taxonomy in 1866 by Ernst Haeckel, leaving plants, animals, and the protists (Protista), defined as a "kingdom of primitive forms".

In 1938, Herbert Copeland resurrected Hogg's label, arguing that Haeckel's term Protista included anucleated microbes such as bacteria, which the term "Protoctista" (literally meaning "first established beings") did not. In contrast, Copeland's term included nucleated eukaryotes such as diatoms, green algae and fungi. This classification was the basis for Whittaker's later definition of Fungi, Animalia, Plantae and Protista as the four kingdoms of life. The kingdom Protista was later modified to separate prokaryotes into the separate kingdom of Monera, leaving the protists as a group of eukaryotic microorganisms. These five kingdoms remained the accepted classification until the development of molecular phylogenetics in the late 20th century, when it became apparent that neither protists nor monera were single groups of related organisms (they were not monophyletic groups).

Modern Classifications

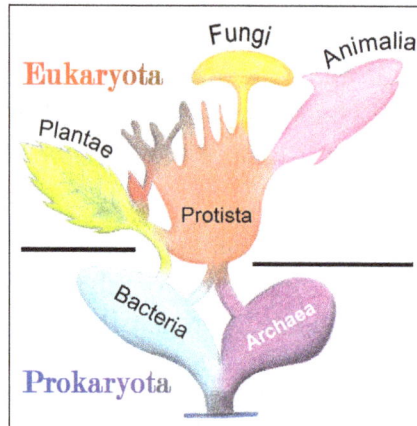

Phylogenetic and symbiogenetic tree of living organisms, showing the origins of eukaryotes.

Systematists today do not treat Protista as a formal taxon, but the term "protist" is still commonly used for convenience in two ways. The most popular contemporary definition is a phylogenetic one, that identifies a paraphyletic group: a protist is any eukaryote that is not an animal, (land) plant, or (true) fungus; this definition excludes many unicellular groups, like the Microsporidia (fungi), many Chytridiomycetes (fungi), and yeasts (fungi), and also a non-unicellular group included in Protista in the past, the Myxozoa (animal). Some systematists[who?] judge paraphyletic taxa acceptable, and use Protista in this sense as a formal taxon (as found in some secondary textbooks, for pedagogical purpose).

The other definition describes protists primarily by functional or biological criteria: protists are essentially those eukaryotes that are never multicellular, that either exist as independent cells, or if they occur in colonies, do not show differentiation into tissues (but vegetative cell differentiation may occur restricted to sexual reproduction, alternate vegetative morphology, and quiescent or resistant stages, such as cysts); this definition excludes many brown, multicellular red and green algae, which may have tissues.

The taxonomy of protists is still changing. Newer classifications attempt to present monophyletic groups based on morphological (especially ultrastructural), biochemical (chemotaxonomy) and DNA sequence (molecular research) information. However, there are sometimes discordances between molecular and morphological investigations; these can be categorized as two types: (i) one morphology, multiple lineages (e.g. morphological convergence, cryptic species) and (ii) one lineage, multiple morphologies (e.g. phenotypic plasticity, multiple life-cycle stages).

Because the protists as a whole are paraphyletic, new systems often split up or abandon the kingdom, instead treating the protist groups as separate lines of eukaryotes. The recent scheme by Adl et al. does not recognize formal ranks (phylum, class, etc.) and instead treats groups as clades of phylogenetically related organisms. This is intended to make the classification more stable in the long term and easier to update. Some of the main groups of protists, which may be treated as phyla, are listed in the taxobox, upper right. Many are thought to be monophyletic, though there is still uncertainty. For instance, the Excavata are probably not monophyletic and the chromalveolates are probably only monophyletic if the haptophytes and cryptomonads are excluded.

Metabolism

Nutrition can vary according to the type of protist. Most eukaryotic algae are autotrophic, but the pigments were lost in some groups.[vague] Other protists are heterotrophic, and may present phagotrophy, osmotrophy, saprotrophy or parasitism. Some are mixotrophic. Some protists that do not have / lost chloroplasts/mitochondria have entered into endosymbiontic relationship with other bacteria/algae to replace the missing functionality. For example, Paramecium bursaria and Paulinella have captured a green alga (Zoochlorella) and a cyanobacterium respectively that act as replacements for chloroplast. Meanwhile, a protist, Mixotricha paradoxa that has lost its mitochondria uses endosymbiontic bacteria as mitochondria and ectosymbiontic hair-like bacteria (Treponema spirochetes) for locomotion.

Many protists are flagellate, for example, and filter feeding can take place where flagellates find prey. Other protists can engulf bacteria and other food particles, by extending their cell membrane around them to form a food vacuole and digesting them internally in a process termed phagocytosis.

Table: Nutritional types in protist metabolism.

Nutritional type	Source of energy	Source of carbon	Examples
Photoautotrophs	Sunlight	Organic compounds or carbon fixation	Most algae
Chemoheterotrophs	Organic compounds	Organic compounds	Apicomplexa, Trypanosomes or Amoebae

For most important cellular structures and functions of animal and plants, it can be found a heritage among protists.

Reproduction

Some protists reproduce sexually using gametes, while others reproduce asexually by binary fission.

Some species, for example Plasmodium falciparum, have extremely complex life cycles that involve multiple forms of the organism, some of which reproduce sexually and others asexually. However, it is unclear how frequently sexual reproduction causes genetic exchange between different strains of Plasmodium in nature and most populations of parasitic protists may be clonal lines that rarely exchange genes with other members of their species.

Eukaryotes emerged in evolution more than 1.5 billion years ago. The earliest eukaryotes were likely protists. Although sexual reproduction is widespread among extant eukaryotes, it seemed unlikely until recently, that sex could be a primordial and fundamental characteristic of eukaryotes. A principal reason for this view was that sex appeared to be lacking in certain pathogenic protists whose ancestors branched off early from the eukaryotic family tree. However, several of these protists are now known to be capable of, or to recently have had the capability for, meiosis and hence sexual reproduction. For example, the common intestinal parasite Giardia lamblia was once considered to be a descendant of a protist lineage that predated the emergence of meiosis and sex. However, G. lamblia was recently found to have a core set of genes that function in meiosis and that are widely present among sexual eukaryotes. These results suggested that G. lamblia is capable of meiosis and thus sexual reproduction. Furthermore, direct evidence for meiotic recombination, indicative of sex, was also found in G. lamblia.

The pathogenic parasitic protists of the genus Leishmania have been shown to be capable of a sexual cycle in the invertebrate vector, likened to the meiosis undertaken in the trypanosomes.

Trichomonas vaginalis, a parasitic protist, is not known to undergo meiosis, but when Malik et al. tested for 29 genes that function in meiosis, they found 27 to be present, including 8 of 9 genes specific to meiosis in model eukaryotes. These findings suggest that T. vaginalis may be capable of meiosis. Since 21 of the 29 meiotic genes were also present in G. lamblia, it appears that most of these meiotic genes were likely present in a common ancestor of T. vaginalis and G. lamblia. These two species are descendants of protist lineages that are highly divergent among eukaryotes, leading Malik et al. to suggest that these meiotic genes were likely present in a common ancestor of all eukaryotes.

Based on a phylogenetic analysis, Dacks and Roger proposed that facultative sex was present in the common ancestor of all eukaryotes.

This view was further supported by a study of amoebae by Lahr et al. Amoeba have generally been regarded as asexual protists. However, these authors describe evidence that most amoeboid lineages are anciently sexual, and that the majority of asexual groups likely arose recently and independently. Early researchers (e.g., Calkins) have interpreted phenomena related to chromidia (chromatin granules free in the cytoplasm) in amoeboid organisms as sexual reproduction.

Protists generally reproduce asexually under favorable environmental conditions, but tend to reproduce sexually under stressful conditions, such as starvation or heat shock. Oxidative stress, which is associated with the production of reactive oxygen species leading to DNA damage, also appears to be an important factor in the induction of sex in protists.

Some commonly found Protist pathogens such as Toxoplasma gondii are capable of infecting and undergoing asexual reproduction in a wide variety of animals – which act as secondary or intermediate host – but can undergo sexual reproduction only in the primary or definitive host (for example: felids such as domestic cats in this case).

Ecology

Free-living Protists occupy almost any environment that contains liquid water. Many protists, such as algae, are photosynthetic and are vital primary producers in ecosystems, particularly in the ocean as part of the plankton. Protists make up a large portion of the biomass in both marine and terrestrial environments.

Other protists include pathogenic species, such as the kinetoplastid Trypanosoma brucei, which causes sleeping sickness, and species of the apicomplexan Plasmodium, which cause malaria.

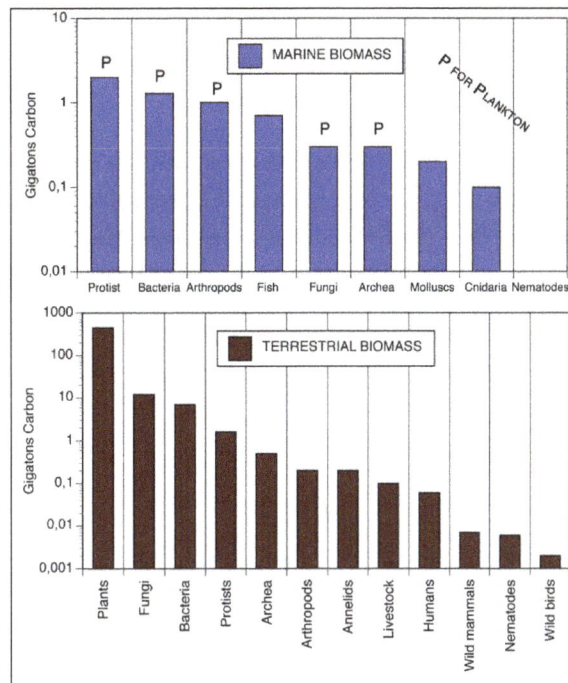

Parasitism: Role as Pathogens

Some protists are significant parasites of animals (e.g.; five species of the parasitic genus Plasmodium cause malaria in humans and many others cause similar diseases in other vertebrates), plants (the oomycete Phytophthora infestans causes late blight in potatoes) or even of other protists. Protist pathogens share many metabolic pathways with their eukaryotic hosts. This makes therapeutic target development extremely difficult – a drug that harms a protist parasite is also likely to harm its animal/plant host. A more thorough understanding of protist biology may allow these diseases to be treated more efficiently. For example, the apicoplast (a nonphotosynthetic chloroplast but essential to carry out important functions other than photosynthesis) present in apicomplexans provides an attractive target for treating diseases caused by dangerous pathogens such as plasmodium.

Recent papers have proposed the use of viruses to treat infections caused by protozoa.

Researchers from the Agricultural Research Service are taking advantage of protists as pathogens to control red imported fire ant (Solenopsis invicta) populations in Argentina. Spore-producing protists such as Kneallhazia solenopsae (recognized as a sister clade or the closest relative to the fungus kingdom now) can reduce red fire ant populations by 53–100%. Researchers have also been able to infect phorid fly parasitoids of the ant with the protist without harming the flies. This turns the flies into a vector that can spread the pathogenic protist between red fire ant colonies.

Fossil Record

Many protists have neither hard parts nor resistant spores, and their fossils are extremely rare or unknown. Examples of such groups include the apicomplexans, most ciliates, some green algae (the Klebsormidiales), choanoflagellates, oomycetes, brown algae, yellow-green algae, Excavata (e.g., euglenids). Some of these have been found preserved in amber (fossilized tree resin) or under unusual conditions (e.g., Paleoleishmania, a kinetoplastid).

Others are relatively common in the fossil record, as the diatoms, golden algae, haptophytes (coccoliths), silicoflagellates, tintinnids (ciliates), dinoflagellates, green algae, red algae, heliozoans, radiolarians, foraminiferans, ebriids and testate amoebae (euglyphids, arcellaceans). Some are even used as paleoecological indicators to reconstruct ancient environments.

More probable eukaryote fossils begin to appear at about 1.8 billion years ago, the acritarchs, spherical fossils of likely algal protists. Another possible representative of early fossil eukaryotes are the Gabonionta.

Fungus

A fungus is any member of the group of eukaryotic organisms that includes microorganisms such as yeasts and molds, as well as the more familiar mushrooms. These organisms are classified as a kingdom, fungi, which is separate from the other eukaryotic life kingdoms of plants and animals.

A characteristic that places fungi in a different kingdom from plants, bacteria, and some protists is chitin in their cell walls. Similar to animals, fungi are heterotrophs; they acquire their food by absorbing dissolved molecules, typically by secreting digestive enzymes into their environment. Fungi do not photosynthesize. Growth is their means of mobility, except for spores (a few of which are flagellated), which may travel through the air or water. Fungi are the principal decomposers in ecological systems. These and other differences place fungi in a single group of related organisms, named the Eumycota (true fungi or Eumycetes), which share a common ancestor (form a monophyletic group), an interpretation that is also strongly supported by molecular phylogenetics. This fungal group is distinct from the structurally similar myxomycetes (slime molds) and oomycetes (water molds). The discipline of biology devoted to the study of fungi is known as mycology. In the past, mycology was regarded as a branch of botany, although it is now known fungi are genetically more closely related to animals than to plants.

Abundant worldwide, most fungi are inconspicuous because of the small size of their structures, and their cryptic lifestyles in soil or on dead matter. Fungi include symbionts of plants, animals, or other fungi and also parasites. They may become noticeable when fruiting, either as mushrooms or as molds. Fungi perform an essential role in the decomposition of organic matter and have fundamental roles in nutrient cycling and exchange in the environment. They have long been used as a direct source of human food, in the form of mushrooms and truffles; as a leavening agent for bread; and in the fermentation of various food products, such as wine, beer, and soy sauce. Since the 1940s, fungi have been used for the production of antibiotics, and, more recently, various enzymes produced by fungi are used industrially and in detergents. Fungi are also used as biological pesticides to control weeds, plant diseases and insect pests. Many species produce bioactive compounds called mycotoxins, such as alkaloids and polyketides, that are toxic to animals including humans. The fruiting structures of a few species contain psychotropic compounds and are consumed recreationally or in traditional spiritual ceremonies. Fungi can break down manufactured materials and buildings, and become significant pathogens of humans and other animals. Losses of crops due to fungal diseases (e.g., rice blast disease) or food spoilage can have a large impact on human food supplies and local economies.

The fungus kingdom encompasses an enormous diversity of taxa with varied ecologies, life cycle strategies, and morphologies ranging from unicellular aquatic chytrids to large mushrooms. However, little is known of the true biodiversity of Kingdom Fungi, which has been estimated at 2.2 million to 3.8 million species. Of these, only about 120,000 have been described, with over 8,000 species known to be detrimental to plants and at least 300 that can be pathogenic to humans. Ever since the pioneering 18th and 19th century taxonomical works of Carl Linnaeus, Christian Hendrik Persoon, and Elias Magnus Fries, fungi have been classified according to their morphology (e.g., characteristics such as spore color or microscopic features) or physiology. Advances in molecular genetics have opened the way for DNA analysis to be incorporated into taxonomy, which has sometimes challenged the historical groupings based on morphology and other traits. Phylogenetic studies published in the last decade have helped reshape the classification within Kingdom Fungi, which is divided into one subkingdom, seven phyla, and ten subphyla.

Characteristics

Before the introduction of molecular methods for phylogenetic analysis, taxonomists considered fungi to be members of the plant kingdom because of similarities in lifestyle: both fungi and plants are mainly immobile, and have similarities in general morphology and growth habitat. Like plants, fungi often grow in soil and, in the case of mushrooms, form conspicuous fruit bodies, which sometimes resemble plants such as mosses. The fungi are now considered a separate kingdom, distinct from both plants and animals, from which they appear to have diverged around one billion years ago (around the start of the Neoproterozoic Era). Some morphological, biochemical, and genetic features are shared with other organisms, while others are unique to the fungi, clearly separating them from the other kingdoms.

Shared Features

- With other eukaryotes: Fungal cells contain membrane-bound nuclei with chromosomes that contain DNA with noncoding regions called introns and coding regions called exons. Fungi have membrane-bound cytoplasmic organelles such as mitochondria, sterol-containing membranes, and ribosomes of the 80S type. They have a characteristic range of soluble carbohydrates and storage compounds, including sugar alcohols (e.g., mannitol), disaccharides, (e.g., trehalose), and polysaccharides (e.g., glycogen, which is also found in animals).

- With animals: Fungi lack chloroplasts and are heterotrophic organisms and so require preformed organic compounds as energy sources.

- With plants: Fungi have a cell wall and vacuoles. They reproduce by both sexual and asexual means, and like basal plant groups (such as ferns and mosses) produce spores. Similar to mosses and algae, fungi typically have haploid nuclei.

- With euglenoids and bacteria: Higher fungi, euglenoids, and some bacteria produce the amino acid L-lysine in specific biosynthesis steps, called the α-aminoadipate pathway.

- The cells of most fungi grow as tubular, elongated, and thread-like (filamentous) structures called hyphae, which may contain multiple nuclei and extend by growing at their tips. Each tip contains a set of aggregated vesicles—cellular structures consisting of proteins, lipids, and other organic molecules—called the Spitzenkörper. Both fungi and oomycetes grow as filamentous hyphal cells. In contrast, similar-looking organisms, such as filamentous green algae, grow by repeated cell division within a chain of cells. There are also single-celled fungi (yeasts) that do not form hyphae, and some fungi have both hyphal and yeast forms.

- In common with some plant and animal species, more than 70 fungal species display bioluminescence.

Unique Features

- Some species grow as unicellular yeasts that reproduce by budding or fission. Dimorphic fungi can switch between a yeast phase and a hyphal phase in response to environmental conditions.

- The fungal cell wall is composed of glucans and chitin; while glucans are also found in plants and chitin in the exoskeleton of arthropods, fungi are the only organisms that combine these two structural molecules in their cell wall. Unlike those of plants and oomycetes, fungal cell walls do not contain cellulose.

Most fungi lack an efficient system for the long-distance transport of water and nutrients, such as the xylem and phloem in many plants. To overcome this limitation, some fungi, such as Armillaria, form rhizomorphs, which resemble and perform functions similar to the roots of plants. As eukaryotes, fungi possess a biosynthetic pathway for producing terpenes that uses mevalonic acid and pyrophosphate as chemical building blocks. Plants and some other organisms have an additional terpene biosynthesis pathway in their chloroplasts, a structure fungi and animals do not have. Fungi produce several secondary metabolites that are similar or identical in structure to those made by plants. Many of the plant and fungal enzymes that make these compounds differ from each other in sequence and other characteristics, which indicates separate origins and convergent evolution of these enzymes in the fungi and plants.

Diversity

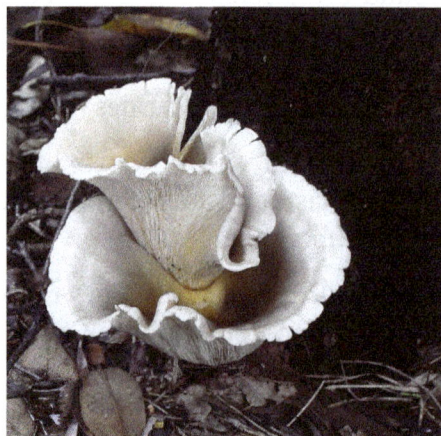

Omphalotus nidiformis, a bioluminescent mushroom.

Fungi have a worldwide distribution, and grow in a wide range of habitats, including extreme environments such as deserts or areas with high salt concentrations or ionizing radiation, as well as in deep sea sediments. Some can survive the intense UV and cosmic radiation encountered during space travel. Most grow in terrestrial environments, though several species live partly or solely in aquatic habitats, such as the chytrid fungus Batrachochytrium dendrobatidis, a parasite that has been responsible for a worldwide decline in amphibian populations. This organism spends part of its life cycle as a motile zoospore, enabling it to propel itself through water and enter its amphibian host. Other examples of aquatic fungi include those living in hydrothermal areas of the ocean.

Bracket fungi on a tree stump.

Around 120,000 species of fungi have been described by taxonomists, but the global biodiversity of the fungus kingdom is not fully understood. A 2017 estimate suggests there may be between 2.2 and 3.8 million species. In mycology, species have historically been distinguished by a variety of methods and concepts. Classification based on morphological characteristics, such as the size and shape of spores or fruiting structures, has traditionally dominated fungal taxonomy. Species may also be distinguished by their biochemical and physiological characteristics, such as their ability to metabolize certain biochemicals, or their reaction to chemical tests. The biological species concept discriminates species based on their ability to mate. The application of molecular tools, such as DNA sequencing and phylogenetic analysis, to study diversity has greatly enhanced the resolution and added robustness to estimates of genetic diversity within various taxonomic groups.

Mycology

Mycology is the branch of biology concerned with the systematic study of fungi, including their genetic and biochemical properties, their taxonomy, and their use to humans as a source of medicine, food, and psychotropic substances consumed for religious purposes, as well as their dangers, such as poisoning or infection. The field of phytopathology, the study of plant diseases, is closely related because many plant pathogens are fungi.

The use of fungi by humans dates back to prehistory; Ötzi the Iceman, a well-preserved mummy of a 5,300-year-old Neolithic man found frozen in the Austrian Alps, carried two species of polypore mushrooms that may have been used as tinder (Fomes fomentarius), or for medicinal purposes (Piptoporus betulinus). Ancient peoples have used fungi as food sources—often unknowingly—for millennia, in the preparation of leavened bread and fermented juices. Some of the oldest written records contain references to the destruction of crops that were probably caused by pathogenic fungi.

Mycology is a relatively new science that became systematic after the development of the microscope in the 17th century. Although fungal spores were first observed by Giambattista della Porta in 1588, the seminal work in the development of mycology is considered to be the publication of Pier Antonio Micheli's 1729 work Nova plantarum genera. Micheli not only observed spores but also showed that, under the proper conditions, they could be induced into growing into the same species of fungi from which they originated. Extending the use of the binomial system of nomenclature introduced by Carl Linnaeus in his Species plantarum, the Dutch Christian Hendrik Persoon established the first classification of mushrooms with such skill as to be considered a founder of modern mycology. Later, Elias Magnus Fries further elaborated the classification of fungi, using spore color and microscopic characteristics, methods still used by taxonomists today. Other notable early contributors to mycology in the 17th–19th and early 20th centuries include Miles Joseph Berkeley, August Carl Joseph Corda, Anton de Bary, the brothers Louis René and Charles Tulasne, Arthur H. R. Buller, Curtis G. Lloyd, and Pier Andrea Saccardo. The 20th century has seen a modernization of mycology that has come from advances in biochemistry, genetics, molecular biology, and biotechnology. The use of DNA sequencing technologies and phylogenetic analysis has provided new insights into fungal relationships and biodiversity, and has challenged traditional morphology-based groupings in fungal taxonomy.

Morphology

Microscopic Structures

Most fungi grow as hyphae, which are cylindrical, thread-like structures 2–10 μm in diameter and up to several centimeters in length. Hyphae grow at their tips (apices); new hyphae are typically formed by emergence of new tips along existing hyphae by a process called branching, or occasionally growing hyphal tips fork, giving rise to two parallel-growing hyphae. Hyphae also sometimes fuse when they come into contact, a process called hyphal fusion (or anastomosis). These growth processes lead to the development of a mycelium, an interconnected network of hyphae. Hyphae can be either septate or coenocytic. Septate hyphae are divided into compartments separated by cross walls (internal cell walls, called septa, that are formed at right angles to the cell wall giving the hypha its shape), with each compartment containing one or more nuclei; coenocytic hyphae are not compartmentalized. Septa have pores that allow cytoplasm, organelles, and sometimes nuclei to pass through; an example is the dolipore septum in fungi of the phylum Basidiomycota. Coenocytic hyphae are in essence multinucleate supercells.

Many species have developed specialized hyphal structures for nutrient uptake from living hosts; examples include haustoria in plant-parasitic species of most fungal phyla, and arbuscules of several mycorrhizal fungi, which penetrate into the host cells to consume nutrients.

Although fungi are opisthokonts—a grouping of evolutionarily related organisms broadly characterized by a single posterior flagellum—all phyla except for the chytrids have lost their posterior flagella. Fungi are unusual among the eukaryotes in having a cell wall that, in addition to glucans (e.g., β-1,3-glucan) and other typical components, also contains the biopolymer chitin.

Macroscopic Structures

Fungal mycelia can become visible to the naked eye, for example, on various surfaces and substrates, such as damp walls and spoiled food, where they are commonly called molds. Mycelia grown on solid agar media in laboratory petri dishes are usually referred to as colonies. These colonies can exhibit growth shapes and colors (due to spores or pigmentation) that can be used as diagnostic features in the identification of species or groups. Some individual fungal colonies can reach extraordinary dimensions and ages as in the case of a clonal colony of Armillaria solidipes, which extends over an area of more than 900 ha (3.5 square miles), with an estimated age of nearly 9,000 years.

Armillaria solidipes

The apothecium—a specialized structure important in sexual reproduction in the ascomycetes—is a cup-shaped fruit body that is often macroscopic and holds the hymenium, a layer of tissue containing the spore-bearing cells. The fruit bodies of the basidiomycetes (basidiocarps) and some ascomycetes can sometimes grow very large, and many are well known as mushrooms.

Growth and Physiology

The growth of fungi as hyphae on or in solid substrates or as single cells in aquatic environments is adapted for the efficient extraction of nutrients, because these growth forms have high surface area to volume ratios. Hyphae are specifically adapted for growth on solid surfaces, and to invade substrates and tissues. They can exert large penetrative mechanical forces; for example, many plant pathogens, including Magnaporthe grisea, form a structure called an appressorium that evolved to puncture plant tissues. The pressure generated by the appressorium, directed against the plant epidermis, can exceed 8 megapascals (1,200 psi). The filamentous fungus Paecilomyces lilacinus uses a similar structure to penetrate the eggs of nematodes.

The mechanical pressure exerted by the appressorium is generated from physiological processes that increase intracellular turgor by producing osmolytes such as glycerol. Adaptations such as these are complemented by hydrolytic enzymes secreted into the environment to digest large organic molecules—such as polysaccharides, proteins, and lipids—into smaller molecules that may then be absorbed as nutrients. The vast majority of filamentous fungi grow in a polar fashion (extending in one direction) by elongation at the tip (apex) of the hypha. Other forms of fungal growth include intercalary extension (longitudinal expansion of hyphal compartments that are below the apex) as in the case of some endophytic fungi, or growth by volume expansion during the development of mushroom stipes and other large organs. Growth of fungi as multicellular structures consisting of somatic and reproductive cells—a feature independently evolved in animals and plants—has several functions, including the development of fruit bodies for dissemination of sexual spores and biofilms for substrate colonization and intercellular communication.

The fungi are traditionally considered heterotrophs, organisms that rely solely on carbon fixed by other organisms for metabolism. Fungi have evolved a high degree of metabolic versatility that allows them to use a diverse range of organic substrates for growth, including simple compounds such as nitrate, ammonia, acetate, or ethanol. In some species the pigment melanin may play a role in extracting energy from ionizing radiation, such as gamma radiation. This form of "radiotrophic" growth has been described for only a few species, the effects on growth rates are small, and the underlying biophysical and biochemical processes are not well known. This process might bear similarity to CO_2 fixation via visible light, but instead uses ionizing radiation as a source of energy.

Reproduction

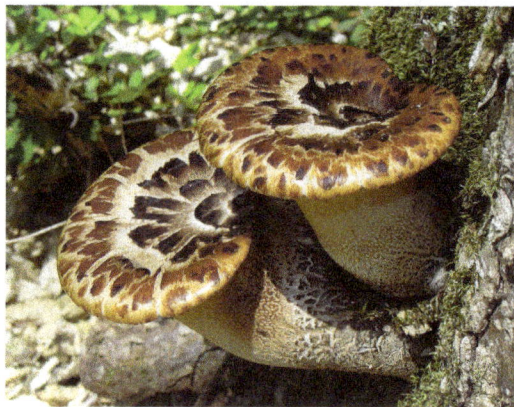

Polyporus squamosus

Fungal reproduction is complex, reflecting the differences in lifestyles and genetic makeup within this diverse kingdom of organisms. It is estimated that a third of all fungi reproduce using more than one method of propagation; for example, reproduction may occur in two well-differentiated stages within the life cycle of a species, the teleomorph and the anamorph. Environmental conditions trigger genetically determined developmental states that lead to the creation of specialized structures for sexual or asexual reproduction. These structures aid reproduction by efficiently dispersing spores or spore-containing propagules.

Asexual Reproduction

Asexual reproduction occurs via vegetative spores (conidia) or through mycelial fragmentation. Mycelial fragmentation occurs when a fungal mycelium separates into pieces, and each component grows into a separate mycelium. Mycelial fragmentation and vegetative spores maintain clonal populations adapted to a specific niche, and allow more rapid dispersal than sexual reproduction. The "Fungi imperfecti" (fungi lacking the perfect or sexual stage) or Deuteromycota comprise all the species that lack an observable sexual cycle. Deuteromycota is not an accepted taxonomic clade, and is now taken to mean simply fungi that lack a known sexual stage.

Sexual Reproduction

Sexual reproduction with meiosis has been directly observed in all fungal phyla except Glomeromycota (genetic analysis suggests meiosis in Glomeromycota as well). It differs in many aspects from sexual reproduction in animals or plants. Differences also exist between fungal groups and can be used to discriminate species by morphological differences in sexual structures and reproductive strategies. Mating experiments between fungal isolates may identify species on the basis of biological species concepts. The major fungal groupings have initially been delineated based on the morphology of their sexual structures and spores; for example, the spore-containing structures, asci and basidia, can be used in the identification of ascomycetes and basidiomycetes, respectively. Fungi employ two mating systems: heterothallic species allow mating only between individuals of opposite mating type, whereas homothallic species can mate, and sexually reproduce, with any other individual or itself.

Most fungi have both a haploid and a diploid stage in their life cycles. In sexually reproducing fungi, compatible individuals may combine by fusing their hyphae together into an interconnected network; this process, anastomosis, is required for the initiation of the sexual cycle. Many ascomycetes and basidiomycetes go through a dikaryotic stage, in which the nuclei inherited from the two parents do not combine immediately after cell fusion, but remain separate in the hyphal cells.

In ascomycetes, dikaryotic hyphae of the hymenium (the spore-bearing tissue layer) form a characteristic hook at the hyphal septum. During cell division, formation of the hook ensures proper distribution of the newly divided nuclei into the apical and basal hyphal compartments. An ascus (plural asci) is then formed, in which karyogamy (nuclear fusion) occurs. Asci are embedded in an ascocarp, or fruiting body. Karyogamy in the asci is followed immediately by meiosis and the production of ascospores. After dispersal, the ascospores may germinate and form a new haploid mycelium.

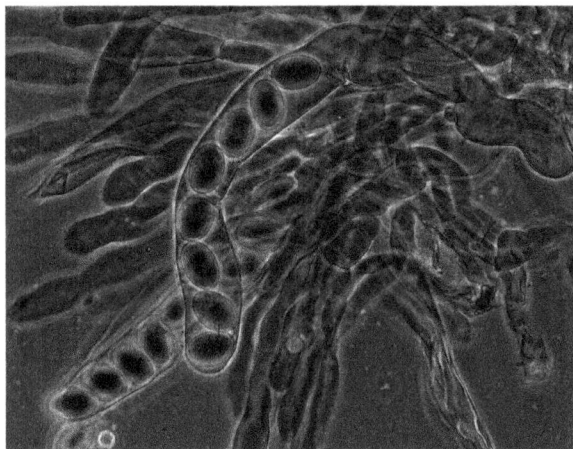

The 8-spore asci of Morchella elata, viewed with phase contrast microscopy.

Sexual reproduction in basidiomycetes is similar to that of the ascomycetes. Compatible haploid hyphae fuse to produce a dikaryotic mycelium. However, the dikaryotic phase is more extensive in the basidiomycetes, often also present in the vegetatively growing mycelium. A specialized anatomical structure, called a clamp connection, is formed at each hyphal septum. As with the structurally similar hook in the ascomycetes, the clamp connection in the basidiomycetes is required for controlled transfer of nuclei during cell division, to maintain the dikaryotic stage with two genetically different nuclei in each hyphal compartment. A basidiocarp is formed in which club-like structures known as basidia generate haploid basidiospores after karyogamy and meiosis. The most commonly known basidiocarps are mushrooms, but they may also take other forms.

In fungi formerly classified as Zygomycota, haploid hyphae of two individuals fuse, forming a gametangium, a specialized cell structure that becomes a fertile gamete-producing cell. The gametangium develops into a zygospore, a thick-walled spore formed by the union of gametes. When the zygospore germinates, it undergoes meiosis, generating new haploid hyphae, which may then form asexual sporangiospores. These sporangiospores allow the fungus to rapidly disperse and germinate into new genetically identical haploid fungal mycelia.

Spore Dispersal

Both asexual and sexual spores or sporangiospores are often actively dispersed by forcible ejection from their reproductive structures. This ejection ensures exit of the spores from the reproductive structures as well as traveling through the air over long distances.

Specialized mechanical and physiological mechanisms, as well as spore surface structures (such as hydrophobins), enable efficient spore ejection. For example, the structure of the spore-bearing cells in some ascomycete species is such that the buildup of substances affecting cell volume and fluid balance enables the explosive discharge of spores into the air. The forcible discharge of single spores termed ballistospores involves formation of a small drop of water (Buller's drop), which upon contact with the spore leads to its projectile release with an initial acceleration of more than 10,000 g; the net result is that the spore is ejected 0.01–0.02 cm, sufficient distance for it to fall through the gills or pores into the air below. Other fungi, like the puffballs, rely on alternative mechanisms for spore release, such as external mechanical forces. The bird's nest fungi use the force of falling water drops to liberate the spores from cup-shaped fruiting bodies. Another strategy is seen in the stinkhorns, a group of fungi with lively colors and putrid odor that attract insects to disperse their spores.

The most common means of spore dispersal is by wind - species using this form of dispersal often produce dry or hydrophobic spores which do not absorb water and are readily scattered by raindrops, for example. Most of the researched species of fungus are transported by wind.

The bird's nest fungus Cyathus stercoreus.

Other Sexual Processes

Besides regular sexual reproduction with meiosis, certain fungi, such as those in the genera Penicillium and Aspergillus, may exchange genetic material via parasexual processes, initiated by anastomosis between hyphae and plasmogamy of fungal cells. The frequency and relative importance of parasexual events is unclear and may be lower than other sexual processes. It is known to play a role in intraspecific hybridization and is likely required for hybridization between species, which has been associated with major events in fungal evolution.

Evolution

In contrast to plants and animals, the early fossil record of the fungi is meager. Factors that likely contribute to the under-representation of fungal species among fossils include the nature of fungal fruiting bodies, which are soft, fleshy, and easily degradable tissues and the microscopic dimensions of most fungal structures, which therefore are not readily evident. Fungal fossils are difficult to distinguish from those of other microbes, and are most easily identified when they resemble extant fungi. Often recovered from a permineralized plant or animal host, these samples are typically studied by making thin-section preparations that can be examined with light microscopy or transmission electron microscopy. Researchers study compression fossils by dissolving the surrounding matrix with acid and then using light or scanning electron microscopy to examine surface details.

The earliest fossils possessing features typical of fungi date to the Paleoproterozoic era, some 2,400 million years ago (Ma); these multicellular benthic organisms had filamentous structures capable of anastomosis. Other studies estimate the arrival of fungal organisms at about 760–1060 Ma on the basis of comparisons of the rate of evolution in closely related groups. For much of the Paleozoic Era (542–251 Ma), the fungi appear to have been aquatic and consisted of organisms similar to the extant chytrids in having flagellum-bearing spores. The evolutionary adaptation from an aquatic to a terrestrial lifestyle necessitated a diversification of ecological strategies for obtaining nutrients, including parasitism, saprobism, and the development of mutualistic relationships such as mycorrhiza and lichenization. Recent studies suggest that the ancestral ecological state of the Ascomycota was saprobism, and that independent lichenization events have occurred multiple times.

In May 2019, scientists reported the discovery of a fossilized fungus, named Ourasphaira giraldae, in the Canadian Arctic, that may have grown on land a billion years ago, well before plants were living on land. Earlier, it had been presumed that the fungi colonized the land during the Cambrian (542–488.3 Ma), also long before land plants. Fossilized hyphae and spores recovered from the Ordovician of Wisconsin (460 Ma) resemble modern-day Glomerales, and existed at a time when the land flora likely consisted of only non-vascular bryophyte-like plants. Prototaxites, which was probably a fungus or lichen, would have been the tallest organism of the late Silurian. Fungal fossils do not become common and uncontroversial until the early Devonian (416–359.2 Ma), when they occur abundantly in the Rhynie chert, mostly as Zygomycota and Chytridiomycota. At about this same time, approximately 400 Ma, the Ascomycota and Basidiomycota diverged, and all modern classes of fungi were present by the Late Carboniferous (Pennsylvanian, 318.1–299 Ma).

Lichen-like fossils have been found in the Doushantuo Formation in southern China dating back to 635–551 Ma. Lichens formed a component of the early terrestrial ecosystems, and the estimated age of the oldest terrestrial lichen fossil is 400 Ma; this date corresponds to the age of the oldest known sporocarp fossil, a Paleopyrenomycites species found in the Rhynie Chert. The oldest fossil with microscopic features resembling modern-day basidiomycetes is Palaeoancistrus, found permineralized with a fern from the Pennsylvanian. Rare in the fossil record are the Homobasidiomycetes (a taxon roughly equivalent to the mushroom-producing species of the Agaricomycetes). Two amber-preserved specimens provide evidence that the earliest known mushroom-forming fungi (the extinct species Archaeomarasmius leggetti) appeared during the late Cretaceous, 90 Ma.

Some time after the Permian–Triassic extinction event (251.4 Ma), a fungal spike (originally thought to be an extraordinary abundance of fungal spores in sediments) formed, suggesting that fungi were the dominant life form at this time, representing nearly 100% of the available fossil record for this period. However, the relative proportion of fungal spores relative to spores formed by algal species is difficult to assess, the spike did not appear worldwide, and in many places it did not fall on the Permian–Triassic boundary.

65 million years ago, immediately after the Cretaceous–Paleogene extinction event that famously killed off most dinosaurs, there is a dramatic increase in evidence of fungi, apparently the death of most plant and animal species leading to a huge fungal bloom like "a massive compost heap".

Taxonomy

Although commonly included in botany curricula and textbooks, fungi are more closely related to animals than to plants and are placed with the animals in the monophyletic group of opisthokonts. Analyses using molecular phylogenetics support a monophyletic origin of fungi. The taxonomy of fungi is in a state of constant flux, especially due to recent research based on DNA comparisons. These current phylogenetic analyses often overturn classifications based on older and sometimes less discriminative methods based on morphological features and biological species concepts obtained from experimental matings.

There is no unique generally accepted system at the higher taxonomic levels and there are frequent name changes at every level, from species upwards. Efforts among researchers are now underway to establish and encourage usage of a unified and more consistent nomenclature. Fungal species can also have multiple scientific names depending on their life cycle and mode (sexual or asexual) of reproduction. Web sites such as Index Fungorum and ITIS list current names of fungal species (with cross-references to older synonyms).

The 2007 classification of Kingdom Fungi is the result of a large-scale collaborative research effort involving dozens of mycologists and other scientists working on fungal taxonomy. It recognizes seven phyla, two of which—the Ascomycota and the Basidiomycota—are contained within a branch representing subkingdom Dikarya, the most species rich and familiar group, including all the mushrooms, most food-spoilage molds, most plant pathogenic fungi, and the beer, wine, and bread yeasts. The accompanying cladogram depicts the major fungal taxa and their relationship to opisthokont and unikont organisms, based on the work of Philippe Silar, "The Mycota: A Comprehensive Treatise on Fungi as Experimental Systems for Basic and Applied Research" and Tedersoo et al. 2018. The lengths of the branches are not proportional to evolutionary distances.

Taxonomic Groups

The major phyla (sometimes called divisions) of fungi have been classified mainly on the basis of characteristics of their sexual reproductive structures. Currently, seven phyla are proposed:

Microsporidia, Chytridiomycota, Blastocladiomycota, Neocallimastigomycota, Glomeromycota, Ascomycota, and Basidiomycota.

Phylogenetic analysis has demonstrated that the Microsporidia, unicellular parasites of animals and protists, are fairly recent and highly derived endobiotic fungi (living within the tissue of another species). One 2006 study concludes that the Microsporidia are a sister group to the true fungi; that is, they are each other's closest evolutionary relative. Hibbett and colleagues suggest that this analysis does not clash with their classification of the Fungi, and although the Microsporidia are elevated to phylum status, it is acknowledged that further analysis is required to clarify evolutionary relationships within this group.

The Chytridiomycota are commonly known as chytrids. These fungi are distributed worldwide. Chytrids and their close relatives Neocallimastigomycota and Blastocladiomycota (below) are the only fungi with active motility, producing zoospores that are capable of active movement through aqueous phases with a single flagellum, leading early taxonomists to classify them as protists. Molecular phylogenies, inferred from rRNA sequences in ribosomes, suggest that the Chytrids are a basal group divergent from the other fungal phyla, consisting of four major clades with suggestive evidence for paraphyly or possibly polyphyly.

The Blastocladiomycota were previously considered a taxonomic clade within the Chytridiomycota. Recent molecular data and ultrastructural characteristics, however, place the Blastocladiomycota as a sister clade to the Zygomycota, Glomeromycota, and Dikarya (Ascomycota and Basidiomycota). The blastocladiomycetes are saprotrophs, feeding on decomposing organic matter, and they are parasites of all eukaryotic groups. Unlike their close relatives, the chytrids, most of which exhibit zygotic meiosis, the blastocladiomycetes undergo sporic meiosis.

The Neocallimastigomycota were earlier placed in the phylum Chytridomycota. Members of this small phylum are anaerobic organisms, living in the digestive system of larger herbivorous mammals and in other terrestrial and aquatic environments enriched in cellulose (e.g., domestic waste landfill sites). They lack mitochondria but contain hydrogenosomes of mitochondrial origin. As in the related chrytrids, neocallimastigomycetes form zoospores that are posteriorly uniflagellate or polyflagellate.

Members of the Glomeromycota form arbuscular mycorrhizae, a form of mutualist symbiosis wherein fungal hyphae invade plant root cells and both species benefit from the resulting increased supply of nutrients. All known Glomeromycota species reproduce asexually. The symbiotic association between the Glomeromycota and plants is ancient, with evidence dating to 400 million years ago. Formerly part of the Zygomycota (commonly known as 'sugar' and 'pin' molds), the Glomeromycota were elevated to phylum status in 2001 and now replace the older phylum Zygomycota. Fungi that were placed in the Zygomycota are now being reassigned to the Glomeromycota, or the subphyla incertae sedis Mucoromycotina, Kickxellomycotina, the Zoopagomycotina and the Entomophthoromycotina. Some well-known examples of fungi formerly in the Zygomycota include black bread mold (Rhizopus stolonifer), and Pilobolus species, capable of ejecting spores several meters through the air. Medically relevant genera include Mucor, Rhizomucor, and Rhizopus.

The Ascomycota, commonly known as sac fungi or ascomycetes, constitute the largest taxonomic group within the Eumycota. These fungi form meiotic spores called ascospores, which are enclosed in a special sac-like structure called an ascus. This phylum includes morels, a few mushrooms and truffles, unicellular yeasts (e.g., of the genera Saccharomyces, Kluyveromyces, Pichia, and Candida), and many filamentous fungi living as saprotrophs, parasites, and mutualistic symbionts (e.g. lichens). Prominent and important genera of filamentous ascomycetes include Aspergillus, Penicillium, Fusarium, and Claviceps. Many ascomycete species have only been observed undergoing asexual reproduction (called anamorphic species), but analysis of molecular data has often been able to identify their closest teleomorphs in the Ascomycota. Because the products of meiosis are retained within the sac-like ascus, ascomycetes have been used for elucidating principles of genetics and heredity (e.g., Neurospora crassa).

Members of the Basidiomycota, commonly known as the club fungi or basidiomycetes, produce meiospores called basidiospores on club-like stalks called basidia. Most common mushrooms belong to this group, as well as rust and smut fungi, which are major pathogens of grains. Other important basidiomycetes include the maize pathogen Ustilago maydis, human commensal species of the genus Malassezia, and the opportunistic human pathogen, Cryptococcus neoformans.

Fungus-Like Organisms

Because of similarities in morphology and lifestyle, the slime molds (mycetozoans, plasmodiophorids, acrasids, Fonticula and labyrinthulids, now in Amoebozoa, Rhizaria, Excavata, Opisthokonta and Stramenopiles, respectively), water molds (oomycetes) and hyphochytrids (both Stramenopiles) were formerly classified in the kingdom Fungi, in groups like Mastigomycotina, Gymnomycota and Phycomycetes. The slime molds were studied also as protozoans, leading to an ambiregnal, duplicated taxonomy.

Unlike true fungi, the cell walls of oomycetes contain cellulose and lack chitin. Hyphochytrids have both chitin and cellulose. Slime molds lack a cell wall during the assimilative phase (except labyrinthulids, which have a wall of scales), and ingest nutrients by ingestion (phagocytosis, except labyrinthulids) rather than absorption (osmotrophy, as fungi, labyrinthulids, oomycetes and hyphochytrids). Neither water molds nor slime molds are closely related to the true fungi, and, therefore, taxonomists no longer group them in the kingdom Fungi. Nonetheless, studies of the oomycetes and myxomycetes are still often included in mycology textbooks and primary research literature.

The Eccrinales and Amoebidiales are opisthokont protists, previously thought to be zygomycete fungi. Other groups now in Opisthokonta (e.g., Corallochytrium, Ichthyosporea) were also at given time classified as fungi. The genus Blastocystis, now in Stramenopiles, was originally classified as a yeast. Ellobiopsis, now in Alveolata, was considered a chytrid. The bacteria were also included in fungi in some classifications, as the group Schizomycetes.

The Rozellida clade, including the "ex-chytrid" Rozella, is a genetically disparate group known mostly from environmental DNA sequences that is a sister group to fungi. Members of the group that have been isolated lack the chitinous cell wall that is characteristic of fungi.

The nucleariids may be the next sister group to the eumycete clade, and as such could be included in an expanded fungal kingdom.

Ecology

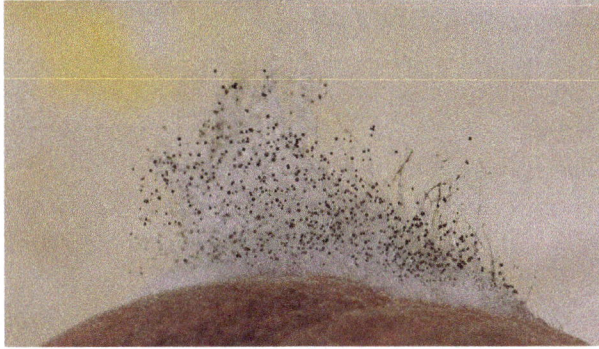

A pin mold decomposing a peach.

Although often inconspicuous, fungi occur in every environment on Earth and play very important roles in most ecosystems. Along with bacteria, fungi are the major decomposers in most terrestrial (and some aquatic) ecosystems, and therefore play a critical role in biogeochemical cycles and in many food webs. As decomposers, they play an essential role in nutrient cycling, especially as saprotrophs and symbionts, degrading organic matter to inorganic molecules, which can then re-enter anabolic metabolic pathways in plants or other organisms.

Symbiosis

Many fungi have important symbiotic relationships with organisms from most if not all Kingdoms. These interactions can be mutualistic or antagonistic in nature, or in the case of commensal fungi are of no apparent benefit or detriment to the host.

With Plants

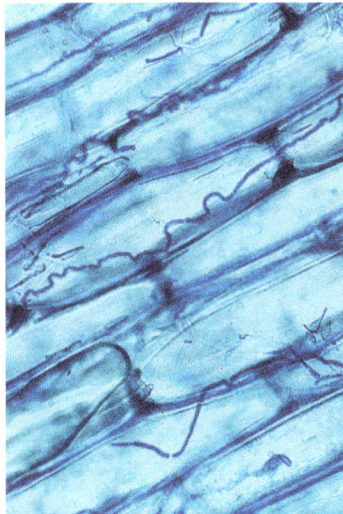

The dark filaments are hyphae of the endophytic fungus Neotyphodium coenophialumin the intercellular spaces of tall fescue leaf sheath tissue.

Mycorrhizal symbiosis between plants and fungi is one of the most well-known plant–fungus associations and is of significant importance for plant growth and persistence in many ecosystems;

over 90% of all plant species engage in mycorrhizal relationships with fungi and are dependent upon this relationship for survival.

The mycorrhizal symbiosis is ancient, dating to at least 400 million years ago. It often increases the plant's uptake of inorganic compounds, such as nitrate and phosphate from soils having low concentrations of these key plant nutrients. The fungal partners may also mediate plant-to-plant transfer of carbohydrates and other nutrients. Such mycorrhizal communities are called "common mycorrhizal networks". A special case of mycorrhiza is myco-heterotrophy, whereby the plant parasitizes the fungus, obtaining all of its nutrients from its fungal symbiont. Some fungal species inhabit the tissues inside roots, stems, and leaves, in which case they are called endophytes. Similar to mycorrhiza, endophytic colonization by fungi may benefit both symbionts; for example, endophytes of grasses impart to their host increased resistance to herbivores and other environmental stresses and receive food and shelter from the plant in return.

With Algae and Cyanobacteria

Lichens are a symbiotic relationship between fungi and photosynthetic algae or cyanobacteria. The photosynthetic partner in the relationship is referred to in lichen terminology as a "photobiont". The fungal part of the relationship is composed mostly of various species of ascomycetes and a few basidiomycetes. Lichens occur in every ecosystem on all continents, play a key role in soil formation and the initiation of biological succession, and are prominent in some extreme environments, including polar, alpine, and semiarid desert regions. They are able to grow on inhospitable surfaces, including bare soil, rocks, tree bark, wood, shells, barnacles and leaves. As in mycorrhizas, the photobiont provides sugars and other carbohydrates via photosynthesis to the fungus, while the fungus provides minerals and water to the photobiont. The functions of both symbiotic organisms are so closely intertwined that they function almost as a single organism; in most cases the resulting organism differs greatly from the individual components. Lichenization is a common mode of nutrition for fungi; around 20% of fungi—between 17,500 and 20,000 described species—are lichenized. Characteristics common to most lichens include obtaining organic carbon by photosynthesis, slow growth, small size, long life, long-lasting (seasonal) vegetative reproductive structures, mineral nutrition obtained largely from airborne sources, and greater tolerance of desiccation than most other photosynthetic organisms in the same habitat.

The lichen Lobaria pulmonaria, a symbiosis of
fungal, algal, and cyanobacterial species.

With Insects

Many insects also engage in mutualistic relationships with fungi. Several groups of ants cultivate fungi in the order Agaricales as their primary food source, while ambrosia beetles cultivate various species of fungi in the bark of trees that they infest. Likewise, females of several wood wasp species (genus Sirex) inject their eggs together with spores of the wood-rotting fungus Amylostereum areolatum into the sapwood of pine trees; the growth of the fungus provides ideal nutritional conditions for the development of the wasp larvae. At least one species of stingless bee has a relationship with a fungus in the genus Monascus, where the larvae consume and depend on fungus transferred from old to new nests. Termites on the African savannah are also known to cultivate fungi, and yeasts of the genera Candida and Lachancea inhabit the gut of a wide range of insects, including neuropterans, beetles, and cockroaches; it is not known whether these fungi benefit their hosts. Fungi ingrowing dead wood are essential for xylophagous insects (e.g. woodboring beetles).[non-primary source needed] They deliver nutrients needed by xylophages to nutritionally scarce dead wood.[non-primary source needed] Thanks to this nutritional enrichment the larvae of woodboring insect is able to grow and develop to adulthood. The larvae of many families of fungicolous flies, particularly those within the superfamily Sciaroidea such as the Mycetophilidae and some Keroplatidae feed on fungal fruiting bodies and sterile mycorrhizae.

As Pathogens and Parasites

Many fungi are parasites on plants, animals (including humans), and other fungi. Serious pathogens of many cultivated plants causing extensive damage and losses to agriculture and forestry include the rice blast fungus Magnaporthe oryzae, tree pathogens such as Ophiostoma ulmi and Ophiostoma novo-ulmi causing Dutch elm disease and Cryphonectria parasitica responsible for chestnut blight, and plant pathogens in the genera Fusarium, Ustilago, Alternaria, and Cochliobolus. Some carnivorous fungi, like Paecilomyces lilacinus, are predators of nematodes, which they capture using an array of specialized structures such as constricting rings or adhesive nets. Many fungi that are plant pathogens, such as Magnaporthe oryzae, can switch from being biotrophic (parasitic on living plants) to being necrotrophic (feeding on the dead tissues of plants they have killed). This same principle is applied to fungi-feeding parasites, including Asterotremella albida, which feeds on the fruit bodies of other fungi both while they are living and after they are dead.

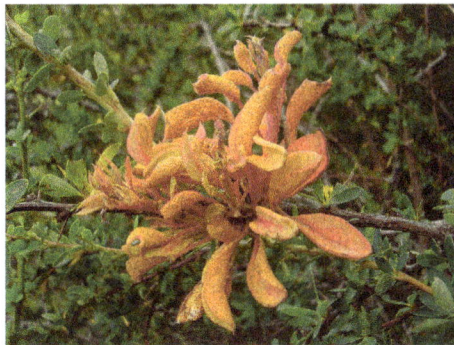

The plant pathogen Aecidium magellanicum causes
calafate rust, seen here on a Berberis shrub in Chile.

Some fungi can cause serious diseases in humans, several of which may be fatal if untreated. These include aspergillosis, candidiasis, coccidioidomycosis, cryptococcosis, histoplasmosis, mycetomas, and paracoccidioidomycosis. Furthermore, persons with immuno-deficiencies are particularly susceptible to disease by genera such as Aspergillus, Candida, Cryptoccocus, Histoplasma, and Pneumocystis. Other fungi can attack eyes, nails, hair, and especially skin, the so-called dermatophytic and keratinophilic fungi, and cause local infections such as ringworm and athlete's foot. Fungal spores are also a cause of allergies, and fungi from different taxonomic groups can evoke allergic reactions.

As Targets of Mycoparasites

The organisms which parasitize fungi are known as mycoparasitic organisms. Certain species of the genus Pythium, which are oomycetes, have potential as biocontrol agents against certain fungi. Fungi can also act as mycoparasites or antagonists of other fungi, such as Hypomyces chrysospermus, which grows on bolete mushrooms.

Mycotoxins

Many fungi produce biologically active compounds, several of which are toxic to animals or plants and are therefore called mycotoxins. Of particular relevance to humans are mycotoxins produced by molds causing food spoilage, and poisonous mushrooms. Particularly infamous are the lethal amatoxins in some Amanita mushrooms, and ergot alkaloids, which have a long history of causing serious epidemics of ergotism (St Anthony's Fire) in people consuming rye or related cereals contaminated with sclerotia of the ergot fungus, Claviceps purpurea. Other notable mycotoxins include the aflatoxins, which are insidious liver toxins and highly carcinogenic metabolites produced by certain Aspergillus species often growing in or on grains and nuts consumed by humans, ochratoxins, patulin, and trichothecenes (e.g., T-2 mycotoxin) and fumonisins, which have significant impact on human food supplies or animal livestock.

Mycotoxins are secondary metabolites (or natural products), and research has established the existence of biochemical pathways solely for the purpose of producing mycotoxins and other natural products in fungi. Mycotoxins may provide fitness benefits in terms of physiological adaptation, competition with other microbes and fungi, and protection from consumption (fungivory). Many fungal secondary metabolites (or derivatives) are used medically.

Pathogenic Mechanisms

Ustilago maydis is a pathogenic plant fungus that causes smut disease in maize and teosinte. Plants have evolved efficient defense systems against pathogenic microbes such as U. maydis. A rapid defense reaction after pathogen attack is the oxidative burst where the plant produces reactive oxygen species at the site of the attempted invasion. U. maydis can respond to the oxidative burst with an oxidative stress response, regulated by the gene YAP1. The response protects U. maydis from the host defense, and is necessary for the pathogen's virulence. Furthermore, U. maydis has a well-established recombinational DNA repair system which acts during mitosis and meiosis. The system may assist the pathogen in surviving DNA damage arising from the host plant's oxidative defensive response to infection.

Cryptococcus neoformans is an encapsulated yeast that can live in both plants and animals. C. neoformans usually infects the lungs, where it is phagocytosed by alveolar macrophages. Some C.

neoformans can survive inside macrophages, which appears to be the basis for latency, disseminated disease, and resistance to antifungal agents. One mechanism by which C. neoformans survives the hostile macrophage environment is by up-regulating the expression of genes involved in the oxidative stress response. Another mechanism involves meiosis. The majority of C. neoformans are mating "type a". Filaments of mating "type a" ordinarily have haploid nuclei, but they can become diploid (perhaps by endoduplication or by stimulated nuclear fusion) to form blastospores. The diploid nuclei of blastospores can undergo meiosis, including recombination, to form haploid basidiospores that can be dispersed. This process is referred to as monokaryotic fruiting. this process requires a gene called DMC1, which is a conserved homologue of genes recA in bacteria and RAD51 in eukaryotes, that mediates homologous chromosome pairing during meiosis and repair of DNA double-strand breaks. Thus, C. neoformans can undergo a meiosis, monokaryotic fruiting, that promotes recombinational repair in the oxidative, DNA damaging environment of the host macrophage, and the repair capability may contribute to its virulence.

Human Use

The human use of fungi for food preparation or preservation and other purposes is extensive and has a long history. Mushroom farming and mushroom gathering are large industries in many countries. The study of the historical uses and sociological impact of fungi is known as ethnomycology. Because of the capacity of this group to produce an enormous range of natural products with antimicrobial or other biological activities, many species have long been used or are being developed for industrial production of antibiotics, vitamins, and anti-cancer and cholesterol-lowering drugs. More recently, methods have been developed for genetic engineering of fungi, enabling metabolic engineering of fungal species. For example, genetic modification of yeast species—which are easy to grow at fast rates in large fermentation vessels—has opened up ways of pharmaceutical production that are potentially more efficient than production by the original source organisms.

Therapeutic Uses

Modern Chemotherapeutics

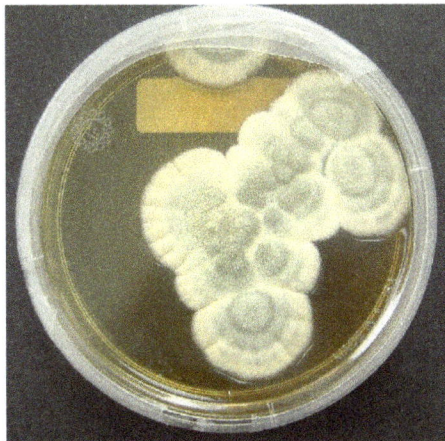

The mould Penicillium chrysogenum was the source of penicillin G.

Many species produce metabolites that are major sources of pharmacologically active drugs. Particularly important are the antibiotics, including the penicillins, a structurally related group of β-lactam antibiotics that are synthesized from small peptides. Although naturally occurring penicillins such as penicillin G (produced by Penicillium chrysogenum) have a relatively narrow spectrum of biological activity, a wide range of other penicillins can be produced by chemical modification of the natural penicillins. Modern penicillins are semisynthetic compounds, obtained initially from fermentation cultures, but then structurally altered for specific desirable properties. Other antibiotics produced by fungi include: ciclosporin, commonly used as an immunosuppressant during transplant surgery; and fusidic acid, used to help control infection from methicillin-resistant Staphylococcus aureus bacteria. Widespread use of antibiotics for the treatment of bacterial diseases, such as tuberculosis, syphilis, leprosy, and others began in the early 20th century and continues to date. In nature, antibiotics of fungal or bacterial origin appear to play a dual role: at high concentrations they act as chemical defense against competition with other microorganisms in species-rich environments, such as the rhizosphere, and at low concentrations as quorum-sensing molecules for intra- or interspecies signaling. Other drugs produced by fungi include griseofulvin isolated from Penicillium griseofulvum, used to treat fungal infections, and statins (HMG-CoA reductase inhibitors), used to inhibit cholesterol synthesis. Examples of statins found in fungi include mevastatin from Penicillium citrinum and lovastatin from Aspergillus terreus and the oyster mushroom. Fungi produce compounds that inhibit viruses and cancer cells. Specific metabolites, such as polysaccharide-K, ergotamine, and β-lactam antibiotics, are routinely used in clinical medicine. The shiitake mushroom is a source of lentinan, a clinical drug approved for use in cancer treatments in several countries, including Japan. In Europe and Japan, polysaccharide-K (brand name Krestin), a chemical derived from Trametes versicolor, is an approved adjuvant for cancer therapy.

Traditional and Folk Medicine

Certain mushrooms enjoy usage as therapeutics in folk medicines, such as Traditional Chinese medicine. Notable medicinal mushrooms with a well-documented history of use include Agaricus subrufescens, Ganoderma lucidum, Psilocybe and Ophiocordyceps sinensis.

Cultured Foods

Baker's yeast or Saccharomyces cerevisiae, a unicellular fungus, is used to make bread and other wheat-based products, such as pizza dough and dumplings. Yeast species of the genus Saccharomyces are also used to produce alcoholic beverages through fermentation. Shoyu koji mold (Aspergillus oryzae) is an essential ingredient in brewing Shoyu (soy sauce) and sake, and the preparation of miso, while Rhizopus species are used for making tempeh. Several of these fungi are domesticated species that were bred or selected according to their capacity to ferment food without producing harmful mycotoxins, which are produced by very closely related Aspergilli. Quorn, a meat substitute, is made from Fusarium venenatum.

In Food

Edible mushrooms include commercially raised and wild-harvested fungi. Agaricus bisporus, sold as button mushrooms when small or Portobello mushrooms when larger, is the most widely cultivated species in the West, used in salads, soups, and many other dishes. Many Asian fungi are commercially grown and have increased in popularity in the West. They are often available fresh in grocery stores and markets, including straw mushrooms (Volvariella volvacea), oyster mushrooms (Pleurotus ostreatus), shiitakes (Lentinula edodes), and enokitake (Flammulina spp.).

Many other mushroom species are harvested from the wild for personal consumption or commercial sale. Milk mushrooms, morels, chanterelles, truffles, black trumpets, and porcini mushrooms (Boletus edulis) (also known as king boletes) demand a high price on the market. They are often used in gourmet dishes.

Certain types of cheeses require inoculation of milk curds with fungal species that impart a unique flavor and texture to the cheese. Examples include the blue color in cheeses such as Stilton or Roquefort, which are made by inoculation with Penicillium roqueforti. Molds used in cheese production are non-toxic and are thus safe for human consumption; however, mycotoxins (e.g., aflatoxins, roquefortine C, patulin, or others) may accumulate because of growth of other fungi during cheese ripening or storage.

A selection of edible mushrooms eaten in Asia.

Stilton cheese veined with Penicillium roqueforti.

Poisonous Fungi

Amanita phalloides accounts for the majority of fatal mushroom poisonings worldwide. It sometimes lacks the greenish color seen here.

Amanita phalloides accounts for the majority of fatal mushroom poisonings worldwide. It sometimes lacks the greenish color seen here.

Many mushroom species are poisonous to humans and cause a range of reactions including slight digestive problems, allergic reactions, hallucinations, severe organ failure, and death. Genera with mushrooms containing deadly toxins include Conocybe, Galerina, Lepiota, and, the most infamous, Amanita. The latter genus includes the destroying angel (A. virosa) and the death cap (A. phalloides), the most common cause of deadly mushroom poisoning. The false morel (Gyromitra esculenta) is occasionally considered a delicacy when cooked, yet can be highly toxic when eaten raw. Tricholoma equestre was considered edible until it was implicated in serious poisonings causing rhabdomyolysis. Fly agaric mushrooms (Amanita muscaria) also cause occasional non-fatal poisonings, mostly as a result of ingestion for its hallucinogenic properties. Historically, fly agaric was used by different peoples in Europe and Asia and its present usage for religious or shamanic purposes is reported from some ethnic groups such as the Koryak people of north-eastern Siberia.

As it is difficult to accurately identify a safe mushroom without proper training and knowledge, it is often advised to assume that a wild mushroom is poisonous and not to consume it.

Pest Control

In agriculture, fungi may be useful if they actively compete for nutrients and space with pathogenic microorganisms such as bacteria or other fungi via the competitive exclusion principle, or if they are parasites of these pathogens. For example, certain species may be used to eliminate or suppress the growth of harmful plant pathogens, such as insects, mites, weeds, nematodes, and other fungi that cause diseases of important crop plants. This has generated strong interest in practical applications that use these fungi in the biological control of these agricultural pests. Entomopathogenic fungi can be used as biopesticides, as they actively kill insects. Examples that have been used as biological insecticides are Beauveria bassiana, Metarhizium spp, Hirsutella spp, Paecilomyces (Isaria) spp, and Lecanicillium lecanii. Endophytic fungi of grasses of the genus Neotyphodium, such as N. coenophialum, produce alkaloids that are toxic to a range of invertebrate and vertebrate herbivores. These alkaloids protect grass plants from herbivory, but several endophyte alkaloids can poison grazing animals, such as cattle and sheep. Infecting cultivars of pasture or forage grasses with Neotyphodium endophytes is one approach being used in grass breeding programs; the fungal strains are selected for producing only alkaloids that increase resistance to herbivores such as insects, while being non-toxic to livestock.

Grasshoppers killed by Beauveria bassiana.

Bioremediation

Certain fungi, in particular white-rot fungi, can degrade insecticides, herbicides, pentachlorophenol, creosote, coal tars, and heavy fuels and turn them into carbon dioxide, water, and basic elements. Fungi have been shown to biomineralize uranium oxides, suggesting they may have application in the bioremediation of radioactively polluted sites.

Model Organisms

Several pivotal discoveries in biology were made by researchers using fungi as model organisms, that is, fungi that grow and sexually reproduce rapidly in the laboratory. For example, the one gene-one enzyme hypothesis was formulated by scientists using the bread mold Neurospora crassa to test their biochemical theories. Other important model fungi are Aspergillus nidulans and the yeasts Saccharomyces cerevisiae and Schizosaccharomyces pombe, each of which with a long history of use to investigate issues in eukaryotic cell biology and genetics, such as cell cycle regulation, chromatin structure, and gene regulation. Other fungal models have more recently emerged that address specific biological questions relevant to medicine, plant pathology, and industrial uses; examples include Candida albicans, a dimorphic, opportunistic human pathogen, Magnaporthe grisea, a plant pathogen, and Pichia pastoris, a yeast widely used for eukaryotic protein production.

Plant

Plants are mainly multicellular, predominantly photosynthetic eukaryotes of the kingdom Plantae. Historically, plants were treated as one of two kingdoms including all living things that were not animals, and all algae and fungi were treated as plants. However, all current definitions of Plantae exclude the fungi and some algae, as well as the prokaryotes (the archaea and bacteria). By one definition, plants form the clade Viridiplantae, a group that includes the flowering plants, conifers and other gymnosperms, ferns and their allies, hornworts, liverworts, mosses and the green algae, but excludes the red and brown algae.

Green plants obtain most of their energy from sunlight via photosynthesis by primary chloroplasts that are derived from endosymbiosis with cyanobacteria. Their chloroplasts contain chlorophylls a and b, which gives them their green color. Some plants are parasitic or mycotrophic and have lost the ability to produce normal amounts of chlorophyll or to photosynthesize. Plants are characterized by sexual reproduction and alternation of generations, although asexual reproduction is also common.

There are about 320 thousand species of plants, of which the great majority, some 260–290 thousand, are seed plants. Green plants provide a substantial proportion of the world's molecular oxygen and are the basis of most of Earth's ecosystems, especially on land. Plants that produce grain, fruit and vegetables form humankind's basic foods, and have been domesticated for millennia. Plants have many cultural and other uses, as ornaments, building materials, writing material and, in great variety, they have been the source of medicines and psychoactive drugs. The scientific study of plants is known as botany, a branch of biology.

All living things were traditionally placed into one of two groups, plants and animals. This classification may date from Aristotle (384 BC – 322 BC), who made the distincton between plants, which generally do not move, and animals, which often are mobile to catch their food.

Much later, when Linnaeus created the basis of the modern system of scientific classification, these two groups became the kingdoms Vegetabilia (later Metaphyta or Plantae) and Animalia (also called Metazoa). Since then, it has become clear that the plant kingdom as originally defined included several unrelated groups, and the fungi and several groups of algae were removed to new kingdoms. However, these organisms are still often considered plants, particularly in popular contexts.

The term "plant" generally implies the possession of the following traits: multicellularity, possession of cell walls containing cellulose, and the ability to carry out photosynthesis with primary chloroplasts.

Current Definitions of Plantae

When the name Plantae or plant is applied to a specific group of organisms or taxon, it usually refers to one of four concepts. From least to most inclusive, these four groupings are:

Name(s)	Scope	Description
Land plants, also known as Embryophyta	Plantae sensu strictissimo	Plants in the strictest sense include the liverworts, hornworts, mosses, and vascular plants, as well as fossil plants similar to these surviving groups.
Green plants, also known as Viridiplantae, Viridiphyta, Chlorobionta or Chloroplastida	Plantae sensu stricto	Plants in a strict sense include the green algae, and land plants that emerged within them, including stoneworts. The relationships between plant groups are still being worked out, and the names given to them vary considerably. The clade Viridiplantae encompasses a group of organisms that have cellulose in their cell walls, possess chlorophylls a and b and have plastids bound by only two membranes that are capable of photosynthesis and of storing starch.
Archaeplastida, also known as Plastida or Primoplantae	Plantae sensu lato	Plants in a broad sense comprise the green plants listed above plus the red algae (Rhodophyta) and the glaucophyte algae (Glaucophyta that store Floridean starch outside the plastids, in the cytoplasm. This clade includes all of the organisms that eons ago acquired their primary chloroplasts directly by engulfing cyanobacteria.
Old definitions of plant (obsolete)	Plantae sensu amplo	Plants in the widest sense refers to older, obsolete classifications that placed diverse algae, fungi or bacteria in Plantae.

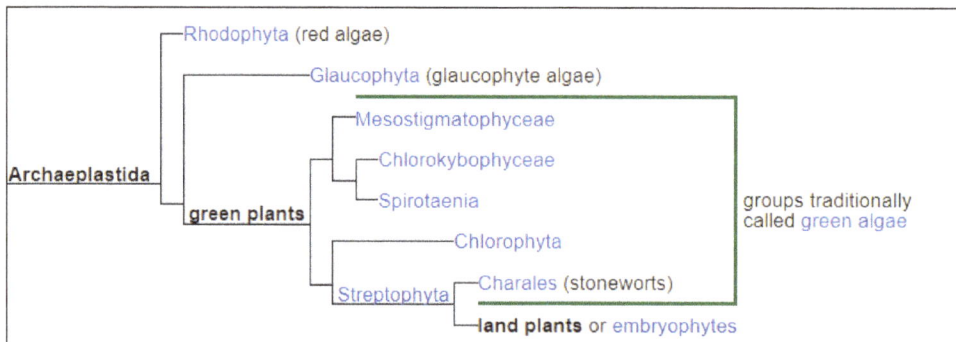

Another way of looking at the relationships between the different groups that have been called "plants" is through a cladogram, which shows their evolutionary relationships. These are not yet completely settled, but one accepted relationship between the three groups described above is shown below.

The way in which the groups of green algae are combined and named varies considerably between authors.

Algae

Algae comprise several different groups of organisms which produce food by photosynthesis and thus have traditionally been included in the plant kingdom. The seaweeds range from large multicellular algae to single-celled organisms and are classified into three groups, the green algae, red algae and brown algae. There is good evidence that the brown algae evolved independently from the others, from non-photosynthetic ancestors that formed endosymbiotic relationships with red algae rather than from cyanobacteria, and they are no longer classified as plants as defined here.

The Viridiplantae, the green plants – green algae and land plants – form a clade, a group consisting of all the descendants of a common ancestor. With a few exceptions, the green plants have the following features in common; primary chloroplasts derived from cyanobacteria containing chlorophylls a and b, cell walls containing cellulose, and food stores in the form of starch contained within the plastids. They undergo closed mitosis without centrioles, and typically have mitochondria with flat cristae. The chloroplasts of green plants are surrounded by two membranes, suggesting they originated directly from endosymbiotic cyanobacteria.

Two additional groups, the Rhodophyta (red algae) and Glaucophyta (glaucophyte algae), also have primary chloroplasts that appear to be derived directly from endosymbiotic cyanobacteria, although they differ from Viridiplantae in the pigments which are used in photosynthesis and so are different in colour. These groups also differ from green plants in that the storage polysaccharide is floridean starch and is stored in the cytoplasm rather than in the plastids. They appear to have had a common origin with Viridiplantae and the three groups form the clade Archaeplastida, whose name implies that their chloroplasts were derived from a single ancient endosymbiotic event. This is the broadest modern definition of the term 'plant'.

In contrast, most other algae (e.g. brown algae/diatoms, haptophytes, dinoflagellates, and euglenids) not only have different pigments but also have chloroplasts with three or four surrounding membranes. They are not close relatives of the Archaeplastida, presumably having acquired chloroplasts separately from ingested or symbiotic green and red algae. They are thus not included in even the broadest modern definition of the plant kingdom, although they were in the past.

The green plants or Viridiplantae were traditionally divided into the green algae (including the stoneworts) and the land plants. However, it is now known that the land plants evolved from within a group of green algae, so that the green algae by themselves are a paraphyletic group, i.e. a group that excludes some of the descendants of a common ancestor. Paraphyletic groups are generally avoided in modern classifications, so that in recent treatments the Viridiplantae have been divided into two clades, the Chlorophyta and the Streptophyta (including the land plants and Charophyta).

The Chlorophyta (a name that has also been used for all green algae) are the sister group to the Charophytes, from which the land plants evolved. There are about 4,300 species, mainly unicellular or multicellular marine organisms such as the sea lettuce, Ulva.

The other group within the Viridiplantae are the mainly freshwater or terrestrial Streptophyta, which consists of the land plants together with the Charophyta, itself consisting of several groups of green algae such as the desmids and stoneworts. Streptophyte algae are either unicellular or form multicellular filaments, branched or unbranched. The genus Spirogyra is a filamentous streptophyte alga familiar to many, as it is often used in teaching and is one of the organisms responsible for the algal "scum" on ponds. The freshwater stoneworts strongly resemble land plants and are believed to be their closest relatives. Growing immersed in fresh water, they consist of a central stalk with whorls of branchlets.

Green algae from Ernst Haeckel's Kunstformen der Natur, 1904.

Fungi

Linnaeus' original classification placed the fungi within the Plantae, since they were unquestionably neither animals or minerals and these were the only other alternatives. With 19th century developments in microbiology, Ernst Haeckel introduced the new kingdom Protista in addition to Plantae and Animalia, but whether fungi were best placed in the Plantae or should be reclassified as protists remained controversial. In 1969, Robert Whittaker proposed the creation of the kingdom Fungi. Molecular evidence has since shown that the most recent common ancestor (concestor), of the Fungi was probably more similar to that of the Animalia than to that of Plantae or any other kingdom.

Whittaker's original reclassification was based on the fundamental difference in nutrition between the Fungi and the Plantae. Unlike plants, which generally gain carbon through photosynthesis, and so are called autotrophs, fungi do not possess chloroplasts and generally obtain carbon by breaking down and absorbing surrounding materials, and so are called heterotrophic saprotrophs. In addition, the substructure of multicellular fungi is different from that of plants, taking the form of many chitinous microscopic strands called hyphae, which may be further subdivided into cells or may form a syncytium containing many eukaryotic nuclei. Fruiting bodies, of which mushrooms are the most familiar example, are the reproductive structures of fungi, and are unlike any structures produced by plants.

Diversity

The table below shows some species count estimates of different green plant (Viridiplantae) divisions. It suggests there are about 300,000 species of living Viridiplantae, of which 85–90% are flowering plants. (Note: as these are from different sources and different dates, they are not necessarily comparable, and like all species counts, are subject to a degree of uncertainty in some cases.)

Table:Diversity of living green plant (Viridiplantae) divisions.

Informal group	Division name	Common name	No. of living species	Approximate No. in informal group
Green algae	Chlorophyta	green algae (chloro-phytes)	3,800–4,300	8,500
	Charophyta	green algae (e.g. de-smids & stoneworts)	2,800–6,000	(6,600–10,300)
Bryophytes	Marchantiophyta	liverworts	6,000–8,000	19,000
	Anthocerotophyta	hornworts	100–200	(18,100–20,200)
	Bryophyta	mosses	12,000	
Pteridophytes	Lycopodiophyta	club mosses	1,200	12,000
	Pteridophyta	ferns, whisk ferns & horsetails	11,000	(12,200)
Seed plants	Cycadophyta	cycads	160	260,000
	Ginkgophyta	ginkgo	1	(259,511)
	Pinophyta	conifers	630	
	Gnetophyta	gnetophytes	70	
	Magnoliophyta	flowering plants	258,650	

The naming of plants is governed by the International Code of Nomenclature for algae, fungi, and plants and International Code of Nomenclature for Cultivated Plants.

Evolution

The evolution of plants has resulted in increasing levels of complexity, from the earliest algal mats, through bryophytes, lycopods, ferns to the complex gymnosperms and angiosperms of to-day. Plants in all of these groups continue to thrive, especially in the environments in which they evolved.

An algal scum formed on the land 1,200 million years ago, but it was not until the Ordovician Period, around 450 million years ago, that land plants appeared. However, new evidence from the study of carbon isotope ratios in Precambrian rocks has suggested that complex photosynthetic plants developed on the earth over 1000 m.y.a. For more than a century it has been assumed that the ancestors of land plants evolved in aquatic environments and then adapted to a life on land, an idea usually credited to botanist Frederick Orpen Bower in his 1908 book "The Origin of a Land Flora". A recent alternative view, supported by genetic evidence, is that they evolved from terrestrial single-celled algae, and that even the common ancestor of red and green algae, and the unicellular freshwater algae glaucophytes, originated in a terrestrial environment in freshwater biofilms or microbial mats. Primitive land plants began to diversify in the late Silurian Period, around 420 million years ago, and the results of their diversification are displayed in remarkable detail in an early Devonian fossil assemblage from the Rhynie chert. This chert preserved early plants in cellular detail, petrified in volcanic springs. By the middle of the Devonian Period most of the features recognised in plants today are present, including roots, leaves and secondary wood, and by late Devonian times seeds had evolved. Late Devonian plants had thereby reached a degree of sophistication that allowed them to form forests of tall trees. Evolutionary innovation continued in the Carboniferous and later geological periods and is ongoing today. Most plant groups were relatively unscathed by the Permo-Triassic extinction event, although the structures of communities changed. This may have set the scene for the evolution of flowering plants in the Triassic (~200 million years ago), which exploded in the Cretaceous and Tertiary. The latest major group of plants to evolve were the grasses, which became important in the mid Tertiary, from around 40 million years ago. The grasses, as well as many other groups, evolved new mechanisms of metabolism to survive the low CO_2 and warm, dry conditions of the tropics over the last 10 million years.

A 1997 proposed phylogenetic tree of Plantae, after Kenrick and Crane, is as follows, with modification to the Pteridophyta from Smith et al. The Prasinophyceae are a paraphyletic assemblage of early diverging green algal lineages, but are treated as a group outside the Chlorophyta: later authors have not followed this suggestion.

A newer proposed classification follows Leliaert et al. 2011 and modified with Silar 2016 for the green algae clades and Novíkov & Barabaš-Krasni 2015 for the land plants clade. Notice that the Prasinophyceae are here placed inside the Chlorophyta.

Embryophytes

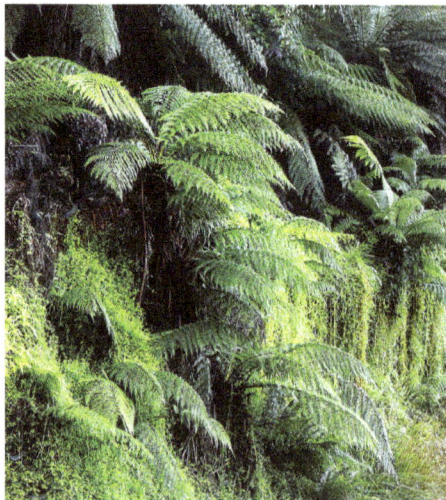

Dicksonia antarctica, a species of tree fern.

The plants that are likely most familiar to us are the multicellular land plants, called embryophytes. Embryophytes include the vascular plants, such as ferns, conifers and flowering plants. They also include the bryophytes, of which mosses and liverworts are the most common.

All of these plants have eukaryotic cells with cell walls composed of cellulose, and most obtain their energy through photosynthesis, using light, water and carbon dioxide to synthesize food. About three hundred plant species do not photosynthesize but are parasites on other species of photo-synthetic plants. Embryophytes are distinguished from green algae, which represent a mode of photosynthetic life similar to the kind modern plants are believed to have evolved from, by having specialized reproductive organs protected by non-reproductive tissues.

Bryophytes first appeared during the early Paleozoic. They mainly live in habitats where moisture is available for significant periods, although some species, such as Targionia, are desiccation-tolerant. Most species of bryophytes remain small throughout their life-cycle. This involves an alternation between two generations: a haploid stage, called the gametophyte, and a diploid stage, called the sporophyte. In bryophytes, the sporophyte is always unbranched and remains nutritionally dependent on its parent gametophyte. The embryophytes have the ability to secrete a cuticle on their outer surface, a waxy layer that confers resistant to des-iccation. In the mosses and hornworts a cuticle is usually only produced on the sporophyte. Stomata are absent from liverworts, but occur on the sporangia of mosses and hornworts, allowing gas exchange.

Vascular plants first appeared during the Silurian period, and by the Devonian had diversified and spread into many different terrestrial environments. They developed a number of adaptations that allowed them to spread into increasingly more arid places, notably the vascular tissues xylem and phloem, that transport water and food throughout the organism. Root systems capable of obtaining soil water and nutrients also evolved during the Devonian. In modern vascular plants, the sporophyte is typically large, branched, nutritionally independent and long-lived, but there is increasing evidence that Paleozoic gametophytes were just as complex as the sporophytes. The gametophytes of all vascular plant groups evolved to become reduced in size and prominence in the life cycle.

In seed plants, the microgametophyte is reduced from a multicellular free-living organism to a few cells in a pollen grain and the miniaturised megagametophyte remains inside the megasporangi-um, attached to and dependent on the parent plant. A megasporangium enclosed in a protective layer called an integument is known as an ovule. After fertilisation by means of sperm produced by pollen grains, an embryo sporophyte develops inside the ovule. The integument becomes a seed coat, and the ovule develops into a seed. Seed plants can survive and reproduce in extremely arid conditions, because they are not dependent on free water for the movement of sperm, or the devel-opment of free living gametophytes.

The first seed plants, pteridosperms (seed ferns), now extinct, appeared in the Devonian and di-versified through the Carboniferous. They were the ancestors of modern gymnosperms, of which four surviving groups are widespread today, particularly the conifers, which are dominant trees in several biomes.

Fossils

Plant fossils include roots, wood, leaves, seeds, fruit, pollen, spores, phytoliths, and amber (the fossilized resin produced by some plants). Fossil land plants are recorded in terrestrial, lacustrine, fluvial and nearshore marine sediments. Pollen, spores and algae (dinoflagellates and acritarchs) are used for dating sedimentary rock sequences. The remains of fossil plants are not as common as fossil animals, although plant fossils are locally abundant in many regions worldwide.

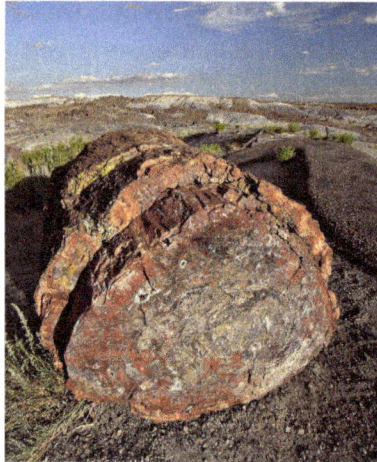

A petrified log in Petrified Forest National Park, Arizona.

The earliest fossils clearly assignable to Kingdom Plantae are fossil green algae from the Cambrian. These fossils resemble calcified multicellular members of the Dasycladales. Earlier Precambrian fossils are known that resemble single-cell green algae, but definitive identity with that group of algae is uncertain.

The earliest fossils attributed to green algae date from the Precambrian (ca. 1200 mya). The resistant outer walls of prasinophyte cysts (known as phycomata) are well preserved in fossil deposits of the Paleozoic (ca. 250–540 mya). A filamentous fossil (Proterocladus) from middle Neoproterozoic deposits (ca. 750 mya) has been attributed to the Cladophorales, while the oldest reliable records of the Bryopsidales, Dasycladales) and stoneworts are from the Paleozoic.

The oldest known fossils of embryophytes date from the Ordovician, though such fossils are fragmentary. By the Silurian, fossils of whole plants are preserved, including the simple vascular plant Cooksonia in mid-Silurian and the much larger and more complex lycophyte Baragwanathia longifolia in late Silurian. From the early Devonian Rhynie chert, detailed fossils of lycophytes and rhyniophytes have been found that show details of the individual cells within the plant organs and the symbiotic association of these plants with fungi of the order Glomales. The Devonian period also saw the evolution of leaves and roots, and the first modern tree, Archaeopteris. This tree with fern-like foliage and a trunk with conifer-like wood was heterosporous producing spores of two different sizes, an early step in the evolution of seeds.

The Coal measures are a major source of Paleozoic plant fossils, with many groups of plants in existence at this time. The spoil heaps of coal mines are the best places to collect; coal itself is the remains of fossilised plants, though structural detail of the plant fossils is rarely visible in coal. In the Fossil Grove at Victoria Park in Glasgow, Scotland, the stumps of Lepidodendron trees are found in their original growth positions.

The fossilized remains of conifer and angiosperm roots, stems and branches may be locally abundant in lake and inshore sedimentary rocks from the Mesozoic and Cenozoic eras. Sequoia and its allies, magnolia, oak, and palms are often found.

Petrified wood is common in some parts of the world, and is most frequently found in arid or desert areas where it is more readily exposed by erosion. Petrified wood is often heavily silicified (the organic material replaced by silicon dioxide), and the impregnated tissue is often preserved in fine detail. Such specimens may be cut and polished using lapidary equipment. Fossil forests of petrified wood have been found in all continents.

Fossils of seed ferns such as Glossopteris are widely distributed throughout several continents of the Southern Hemisphere, a fact that gave support to Alfred Wegener's early ideas regarding Continental drift theory.

Structure, Growth and Development

The leaf is usually the primary site of photosynthesis in plants.

Most of the solid material in a plant is taken from the atmosphere. Through the process of photosynthesis, most plants use the energy in sunlight to convert carbon dioxide from the atmosphere, plus water, into simple sugars. These sugars are then used as building blocks and form the main structural component of the plant. Chlorophyll, a green-colored, magnesium-containing pigment is essential to this process; it is generally present in plant leaves, and often in other plant parts as well. Parasitic plants, on the other hand, use the resources of their host to provide the materials needed for metabolism and growth.

Plants usually rely on soil primarily for support and water (in quantitative terms), but they also obtain compounds of nitrogen, phosphorus, potassium, magnesium and other elemental nutrients from the soil. Epiphytic and lithophytic plants depend on air and nearby debris for nutrients, and carnivorous plants supplement their nutrient requirements, particularly for nitrogen and phosphorus, with insect prey that they capture. For the majority of plants to grow successfully they also require oxygen in the atmosphere and around their roots (soil gas) for respiration. Plants use oxygen and glucose (which may be produced from stored starch) to provide energy. Some plants grow as submerged aquatics, using oxygen dissolved in the surrounding water, and a few specialized vascular plants, such as mangroves and reed (Phragmites australis), can grow with their roots in anoxic conditions.

Factors affecting Growth

The genome of a plant controls its growth. For example, selected varieties or genotypes of wheat grow rapidly, maturing within 110 days, whereas others, in the same environmental conditions, grow more slowly and mature within 155 days.

Growth is also determined by environmental factors, such as temperature, available water, available light, carbon dioxide and available nutrients in the soil. Any change in the availability of these external conditions will be reflected in the plant's growth and the timing of its development.

Biotic factors also affect plant growth. Plants can be so crowded that no single individual produces normal growth, causing etiolation and chlorosis. Optimal plant growth can be hampered by grazing animals, suboptimal soil composition, lack of mycorrhizal fungi, and attacks by insects or plant diseases, including those caused by bacteria, fungi, viruses, and nematodes.

Simple plants like algae may have short life spans as individuals, but their populations are commonly seasonal. Annual plants grow and reproduce within one growing season, biennial plants grow for two growing seasons and usually reproduce in second year, and perennial plants live for many growing seasons and once mature will often reproduce annually. These designations often depend on climate and other environmental factors. Plants that are annual in alpine or temperate regions can be biennial or perennial in warmer climates. Among the vascular plants, perennials include both evergreens that keep their leaves the entire year, and deciduous plants that lose their leaves for some part of it. In temperate and boreal climates, they generally lose their leaves during the winter; many tropical plants lose their leaves during the dry season.

The growth rate of plants is extremely variable. Some mosses grow less than 0.001 millimeters per hour (mm/h), while most trees grow 0.025-0.250 mm/h. Some climbing species, such as kudzu, which do not need to produce thick supportive tissue, may grow up to 12.5 mm/h.

Plants protect themselves from frost and dehydration stress with antifreeze proteins, heat-shock proteins and sugars (sucrose is common). LEA (Late Embryogenesis Abundant) protein expression is induced by stresses and protects other proteins from aggregation as a result of desiccation and freezing.

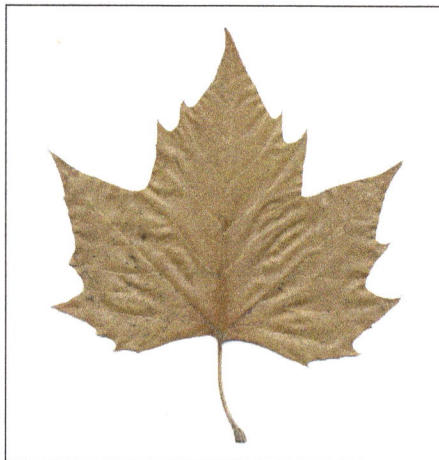

There is no photosynthesis in deciduous leaves in autumn.

Effects of Freezing

When water freezes in plants, the consequences for the plant depend very much on whether the freezing occurs within cells (intracellularly) or outside cells in intercellular spaces. Intracellular freezing, which usually kills the cell regardless of the hardiness of the plant and its tissues, seldom occurs in nature because rates of cooling are rarely high enough to support it. Rates of cooling of several degrees Celsius per minute are typically needed to cause intracellular formation of ice. At rates of cooling of a few degrees Celsius per hour, segregation of ice occurs in intercellular spaces. This may or may not be lethal, depending on the hardiness of the tissue. At freezing temperatures, water in the intercellular spaces of plant tissue freezes first, though the water may remain unfrozen until temperatures drop below −7 °C (19 °F). After the initial formation of intercellular ice, the cells shrink as water is lost to the segregated ice, and the cells undergo freeze-drying. This dehydration is now considered the fundamental cause of freezing injury.

DNA Damage and Repair

Plants are continuously exposed to a range of biotic and abiotic stresses. These stresses often cause DNA damage directly, or indirectly via the generation of reactive oxygen species. Plants are capable of a DNA damage response that is a critical mechanism for maintaining genome stability. The DNA damage response is particularly important during seed germination, since seed quality tends to deteriorate with age in association with DNA damage accumulation. During germination repair processes are activated to deal with this accumulated DNA damage. In particular, single- and double-strand breaks in DNA can be repaired. The DNA checkpoint kinase ATM has a key role in integrating progression through germination with repair responses to the DNA damages accumulated by the aged seed.

Plant Cells

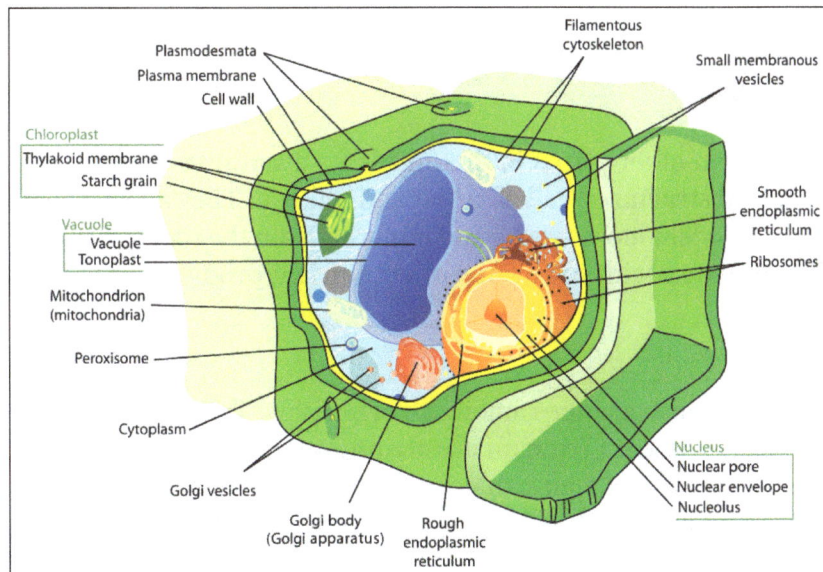

Plant cell structure.

Plant cells are typically distinguished by their large water-filled central vacuole, chloroplasts, and rigid cell walls that are made up of cellulose, hemicellulose, and pectin. Cell division is also characterized by the development of a phragmoplast for the construction of a cell plate in the late stages of cytokinesis. Just as in animals, plant cells differentiate and develop into multiple cell types. Totipotent meristematic cells can differentiate into vascular, storage, protective (e.g. epidermal layer), or reproductive tissues, with more primitive plants lacking some tissue types.

Physiology

Photosynthesis

Plants are photosynthetic, which means that they manufacture their own food molecules using energy obtained from light. The primary mechanism plants have for capturing light energy is the pigment chlorophyll. All green plants contain two forms of chlorophyll, chlorophyll a and chlorophyll b. The latter of these pigments is not found in red or brown algae. The simple equation of photosynthesis is as follows:

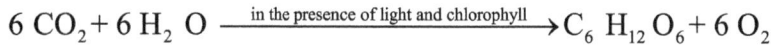

$$6\ CO_2 + 6\ H_2O \xrightarrow{\text{in the presence of light and chlorophyll}} C_6H_{12}O_6 + 6\ O_2$$

Immune System

By means of cells that behave like nerves, plants receive and distribute within their systems information about incident light intensity and quality. Incident light that stimulates a chemical reaction in one leaf, will cause a chain reaction of signals to the entire plant via a type of cell termed a bundle sheath cell. Researchers, from the Warsaw University of Life Sciences in Poland, found that plants have a specific memory for varying light conditions, which prepares their immune systems against seasonal pathogens. Plants use pattern-recognition receptors to recognize conserved microbial signatures. This recognition triggers an immune response. The first plant receptors of conserved microbial signatures were identified in rice (XA21, 1995) and in Arabidopsis thaliana (FLS2, 2000). Plants also carry immune receptors that recognize highly variable pathogen effectors. These include the NBS-LRR class of proteins.

Internal Distribution

Vascular plants differ from other plants in that nutrients are transported between their different parts through specialized structures, called xylem and phloem. They also have roots for taking up water and minerals. The xylem moves water and minerals from the root to the rest of the plant, and the phloem provides the roots with sugars and other nutrient produced by the leaves.

Genomics

Plants have some of the largest genomes among all organisms. The largest plant genome (in terms of gene number) is that of wheat (Triticum asestivum), predicted to encode ≈94,000 genes and thus almost 5 times as many as the human genome. The first plant genome sequenced was that of Arabidopsis thaliana which encodes about 25,500 genes. In terms of sheer DNA sequence, the smallest published genome is that of the carnivorous bladderwort (Utricularia gibba) at 82 Mb (although it still encodes 28,500 genes) while the largest, from the Norway Spruce (Picea abies), extends over 19,600 Mb (encoding about 28,300 genes).

Ecology

The photosynthesis conducted by land plants and algae is the ultimate source of energy and organic material in nearly all ecosystems. Photosynthesis, at first by cyanobacteria and later by photosynthetic eukaryotes, radically changed the composition of the early Earth's anoxic atmosphere, which as a result is now 21% oxygen. Animals and most other organisms are aerobic, relying on oxygen; those that do not are confined to relatively rare anaerobic environments. Plants are the primary producers in most terrestrial ecosystems and form the basis of the food web in those ecosystems. Many animals rely on plants for shelter as well as oxygen and food.

Land plants are key components of the water cycle and several other biogeochemical cycles. Some plants have coevolved with nitrogen fixing bacteria, making plants an important part of the nitrogen cycle. Plant roots play an essential role in soil development and the prevention of soil erosion.

Distribution

Plants are distributed almost worldwide. While they inhabit a multitude of biomes and ecoregions, few can be found beyond the tundras at the northernmost regions of continental shelves. At the southern extremes, plants of the Antarctic flora have adapted tenaciously to the prevailing conditions.

Plants are often the dominant physical and structural component of habitats where they occur. Many of the Earth's biomes are named for the type of vegetation because plants are the dominant organisms in those biomes, such as grasslands, taiga and tropical rainforest.

Ecological Relationships

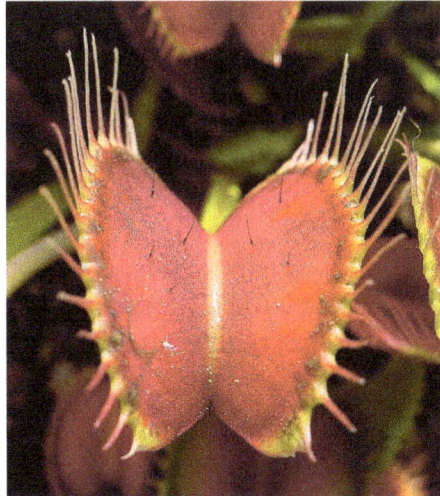

The Venus flytrap, a species of carnivorous plant.

Numerous animals have coevolved with plants. Many animals pollinate flowers in exchange for food in the form of pollen or nectar. Many animals disperse seeds, often by eating fruit and passing the seeds in their feces. Myrmecophytes are plants that have coevolved with ants. The plant provides a home, and sometimes food, for the ants. In exchange, the ants defend the plant from herbivores and sometimes competing plants. Ant wastes provide organic fertilizer.

The majority of plant species have various kinds of fungi associated with their root systems in a kind of mutualistic symbiosis known as mycorrhiza. The fungi help the plants gain water and mineral nutrients from the soil, while the plant gives the fungi carbohydrates manufactured in photosynthesis. Some plants serve as homes for endophytic fungi that protect the plant from herbivores by producing toxins. The fungal endophyte, Neotyphodium coenophialum, in tall fescue (Festuca arundinacea) does tremendous economic damage to the cattle industry in the U.S.

Various forms of parasitism are also fairly common among plants, from the semi-parasitic mistletoe that merely takes some nutrients from its host, but still has photosynthetic leaves, to the fully parasitic broomrape and toothwort that acquire all their nutrients through connections to the roots of other plants, and so have no chlorophyll. Some plants, known as myco-heterotrophs, parasitize mycorrhizal fungi, and hence act as epiparasites on other plants.

Many plants are epiphytes, meaning they grow on other plants, usually trees, without parasitizing them. Epiphytes may indirectly harm their host plant by intercepting mineral nutrients and light that the host would otherwise receive. The weight of large numbers of epiphytes may break tree limbs. Hemiepiphytes like the strangler fig begin as epiphytes but eventually set their own roots and overpower and kill their host. Many orchids, bromeliads, ferns and mosses often grow as epiphytes. Bromeliad epiphytes accumulate water in leaf axils to form phytotelmata that may contain complex aquatic food webs.

Approximately 630 plants are carnivorous, such as the Venus Flytrap (Dionaea muscipula) and sundew (Drosera species). They trap small animals and digest them to obtain mineral nutrients, especially nitrogen and phosphorus.

Importance

The study of plant uses by people is called economic botany or ethnobotany. Human cultivation of plants is part of agriculture, which is the basis of human civilization. Plant agriculture is subdivided into agronomy, horticulture and forestry.

Food

Humans depend on plants for food, either directly or as feed for domestic animals. Agriculture deals with the production of food crops, and has played a key role in the history of world civilizations. Agriculture includes agronomy for arable crops, horticulture for vegetables and fruit, and forestry for timber. About 7,000 species of plant have been used for food, though most of today's food is derived from only 30 species. The major staples include cereals such as rice and wheat, starchy roots and tubers such as cassava and potato, and legumes such as peas and beans. Vegetable oils such as olive oil provide lipids, while fruit and vegetables contribute vitamins and minerals to the diet.

Mechanical harvest of oats.

Medicines

Medicinal plants are a primary source of organic compounds, both for their medicinal and physiological effects, and for the industrial synthesis of a vast array of organic chemicals. Many hundreds of medicines are derived from plants, both traditional medicines used in herbalism and chemical substances purified from plants or first identified in them, sometimes by ethnobotanical search, and then synthesised for use in modern medicine. Modern medicines derived from plants include aspirin, taxol, morphine, quinine, reserpine, colchicine, digitalis and vincristine. Plants used in herbalism include ginkgo, echinacea, feverfew, and Saint John's wort. The pharmacopoeia of Dioscorides, De Materia Medica, describing some 600 medicinal plants, was written between 50 and 70 AD and remained in use in Europe and the Middle East until around 1600 AD; it was the precursor of all modern pharmacopoeias.

Nonfood Products

Plants grown as industrial crops are the source of a wide range of products used in manufacturing, sometimes so intensively as to risk harm to the environment. Nonfood products include essential oils, natural dyes, pigments, waxes, resins, tannins, alkaloids, amber and cork. Products derived from plants include soaps, shampoos, perfumes, cosmetics, paint, varnish, turpentine, rubber, latex, lubricants, linoleum, plastics, inks, and gums. Renewable fuels from plants include firewood, peat and other biofuels. The fossil fuels coal, petroleum and natural gas are derived from the remains of aquatic organisms including phytoplankton in geological time.

Structural resources and fibres from plants are used to construct dwellings and to manufacture clothing. Wood is used not only for buildings, boats, and furniture, but also for smaller items such as musical instruments and sports equipment. Wood is pulped to make paper and cardboard. Cloth is often made from cotton, flax, ramie or synthetic fibres such as rayon and acetate derived from plant cellulose. Thread used to sew cloth likewise comes in large part from cotton.

Timber in storage for later processing at a sawmill

Aesthetic Uses

Thousands of plant species are cultivated for aesthetic purposes as well as to provide shade, modify temperatures, reduce wind, abate noise, provide privacy, and prevent soil erosion. Plants are the basis of a multibillion-dollar per year tourism industry, which includes travel to historic gardens, national parks, rainforests, forests with colorful autumn leaves, and festivals such as Japan's and America's cherry blossom festivals.

While some gardens are planted with food crops, many are planted for aesthetic, ornamental, or conservation purposes. Arboretums and botanical gardens are public collections of living plants. In private outdoor gardens, lawn grasses, shade trees, ornamental trees, shrubs, vines, herbaceous perennials and bedding plants are used. Gardens may cultivate the plants in a naturalistic state, or may sculpture their growth, as with topiary or espalier. Gardening is the most popular leisure activity in the U.S., and working with plants or horticulture therapy is beneficial for rehabilitating people with disabilities.

Plants may also be grown or kept indoors as houseplants, or in specialized buildings such as greenhouses that are designed for the care and cultivation of living plants. Venus Flytrap, sensitive plant and resurrection plant are examples of plants sold as novelties. There are also art forms specializing in the arrangement of cut or living plant, such as bonsai, ikebana, and the arrangement of cut or dried flowers. Ornamental plants have sometimes changed the course of history, as in tulipomania.

Architectural designs resembling plants appear in the capitals of Ancient Egyptian columns, which were carved to resemble either the Egyptian white lotus or the papyrus. Images of plants are often used in painting and photography, as well as on textiles, money, stamps, flags and coats of arms.

A rose espalier at Niedernhall in Germany.

Capitals of ancient Egyptian columns decorated to
resemble papyrus plants.

Scientific and Cultural Uses

Basic biological research has often been done with plants. In genetics, the breeding of pea plants allowed Gregor Mendel to derive the basic laws governing inheritance, and examination of chromosomes in maize allowed Barbara McClintock to demonstrate their connection to inherited traits. The plant Arabidopsis thaliana is used in laboratories as a model organism to understand how genes control the growth and development of plant structures. NASA predicts that space stations or space colonies will one day rely on plants for life support.

Ancient trees are revered and many are famous. Tree rings themselves are an important method of dating in archeology, and serve as a record of past climates.

Plants figure prominently in mythology, religion and literature. They are used as national and state emblems, including state trees and state flowers. Plants are often used as memorials, gifts and to mark special occasions such as births, deaths, weddings and holidays. The arrangement of flowers may be used to send hidden messages.

Negative Effects

Weeds are unwanted plants growing in managed environments such as farms, urban areas, gardens, lawns, and parks. People have spread plants beyond their native ranges and some of these introduced plants become invasive, damaging existing ecosystems by displacing native species, and sometimes becoming serious weeds of cultivation.

Plants may cause harm to animals, including people. Plants that produce windblown pollen invoke allergic reactions in people who suffer from hay fever. A wide variety of plants are poisonous. Toxalbumins are plant poisons fatal to most mammals and act as a serious deterrent to consumption. Several plants cause skin irritations when touched, such as poison ivy. Certain plants contain psychotropic chemicals, which are extracted and ingested or smoked, including nicotine from tobacco, cannabinoids from Cannabis sativa, cocaine from Erythroxylon coca and opium from opium poppy. Smoking causes damage to health or even death, while some drugs may also be harmful or fatal to people. Both illegal and legal drugs derived from plants may have negative effects on the economy, affecting worker productivity and law enforcement costs.

Animal

Animals are multicellular eukaryotic organisms that form the biological kingdom Animalia. With few exceptions, animals consume organic material, breathe oxygen, are able to move, can reproduce sexually, and grow from a hollow sphere of cells, the blastula, during embryonic development. Over 1.5 million living animal species have been described—of which around 1 million are insects—but it has been estimated there are over 7 million animal species in total. Animals range in length from 8.5 millionths of a metre to 33.6 metres (110 ft) and have complex interactions with each other and their environments, forming intricate food webs. The category includes humans, but in colloquial use the term animal often refers only to non-human animals. The study of non-human animals is known as zoology.

Most living animal species are in the Bilateria, a clade whose members have a bilaterally symmetric body plan. The Bilateria include the protostomes—in which many groups of invertebrates are found, such as nematodes, arthropods, and molluscs—and the deuterostomes, containing the echinoderms and chordates (including the vertebrates). Life forms interpreted as early animals were present in the Ediacaran biota of the late Precambrian. Many modern animal phyla became clearly established in the fossil record as marine species during the Cambrian explosion which began around 542 million years ago. 6,331 groups of genes common to all living animals have been identified; these may have arisen from a single common ancestor that lived 650 million years ago.

Aristotle divided animals into those with blood and those without. Carl Linnaeus created the first hierarchical biological classification for animals in 1758 with his Systema Naturae, which Jean-Baptiste Lamarck expanded into 14 phyla by 1809. In 1874, Ernst Haeckel divided the animal kingdom into the multicellular Metazoa (now synonymous with Animalia) and the Protozoa, single-celled organisms no longer considered animals. In modern times, the biological classification of animals relies on advanced techniques, such as molecular phylogenetics, which are effective at demonstrating the evolutionary relationships between animal taxa.

Humans make use of many other animal species for food, including meat, milk, and eggs; for materials, such as leather and wool; as pets; and as working animals for power and transport. Dogs have been used in hunting, while many terrestrial and aquatic animals are hunted for sport. Non-human animals have appeared in art from the earliest times and are featured in mythology and religion.

Characteristics

Animals have several characteristics that set them apart from other living things. Animals are eukaryotic and multicellular, unlike bacteria, which are prokaryotic, and unlike protists, which are eukaryotic but unicellular. Unlike plants and algae, which produce their own nutrients animals are heterotrophic, feeding on organic material and digesting it internally. With very few exceptions, animals breathe oxygen and respire aerobically. All animals are motile (able to spontaneously move their bodies) during at least part of their life cycle, but some animals, such as sponges, corals, mussels, and barnacles, later become sessile. The blastula is a stage in embryonic development that is unique to most animals, allowing cells to be differentiated into specialised tissues and organs.

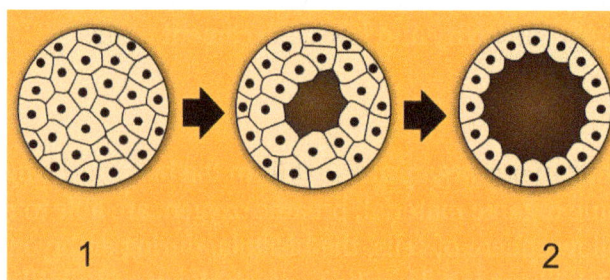

Animals are unique in having the ball of cells of the
early embryo (1) develop into a hollow ball or blastula (2).

Structure

All animals are composed of cells, surrounded by a characteristic extracellular matrix composed of collagen and elastic glycoproteins. During development, the animal extracellular matrix forms a relatively flexible framework upon which cells can move about and be reorganised, making the formation of complex structures possible. This may be calcified, forming structures such as shells, bones, and spicules. In contrast, the cells of other multicellular organisms (primarily algae, plants, and fungi) are held in place by cell walls, and so develop by progressive growth. Animal cells uniquely possess the cell junctions called tight junctions, gap junctions, and desmosomes.

With few exceptions—in particular, the sponges and placozoans—animal bodies are differentiated into tissues. These include muscles, which enable locomotion, and nerve tissues, which transmit signals and coordinate the body. Typically, there is also an internal digestive chamber with either one opening (in Ctenophora, Cnidaria, and flatworms) or two openings (in most bilaterians).

Reproduction and Development

Nearly all animals make use of some form of sexual reproduction. They produce haploid gametes by meiosis; the smaller, motile gametes are spermatozoa and the larger, non-motile gametes are ova. These fuse to form zygotes, which develop via mitosis into a hollow sphere, called a blastula. In sponges, blastula larvae swim to a new location, attach to the seabed, and develop into a new sponge. In most other groups, the blastula undergoes more complicated rearrangement. It first invaginates to form a gastrula with a digestive chamber and two separate germ layers, an external ectoderm and an internal endoderm. In most cases, a third germ layer, the mesoderm, also develops between them. These germ layers then differentiate to form tissues and organs.

Repeated instances of mating with a close relative during sexual reproduction generally leads to inbreeding depression within a population due to the increased prevalence of harmful recessive traits. Animals have evolved numerous mechanisms for avoiding close inbreeding. In some species, such as the splendid fairywren (Malurus splendens), females benefit by mating with multiple males, thus producing more offspring of higher genetic quality.

Some animals are capable of asexual reproduction, which often results in a genetic clone of the parent. This may take place through fragmentation; budding, such as in Hydra and other cnidarians; or parthenogenesis, where fertile eggs are produced without mating, such as in aphids.

Sexual reproduction is nearly universal in animals, such as these dragonflies.

Ecology

Animals are categorised into ecological groups depending on how they obtain or consume organic material, including carnivores, herbivores, omnivores, detritivores, and parasites. Interactions between animals form complex food webs. In carnivorous or omnivorous species, predation is a consumer-resource interaction where a predator feeds on another organism (called its prey). Selective pressures imposed on one another lead to an evolutionary arms race between predator and prey, resulting in various anti-predator adaptations. Almost all multicellular predators are animals. Some consumers use multiple methods; for example, in parasitoid wasps, the larvae feed on the hosts' living tissues, killing them in the process, but the adults primarily consume nectar from flowers. Other animals may have very specific feeding behaviours, such as hawksbill sea turtles that primarily eat sponges.

Most animals rely on the energy produced by plants through photosynthesis. Herbivores eat plant material directly, while carnivores, and other animals on higher trophic levels, typically acquire energy (in the form of reduced carbon) by eating other animals. The carbohydrates, lipids, proteins, and other biomolecules are broken down to allow the animal to grow and to sustain biological processes such as locomotion. Animals living close to hydrothermal vents and cold seeps on the dark sea floor do not depend on the energy of sunlight. Rather, archaea and bacteria in these locations produce organic matter through chemosynthesis (by oxidizing inorganic compounds, such as methane) and form the base of the local food web.

Animals originally evolved in the sea. Lineages of arthropods colonised land around the same time as land plants, probably between 510–471 million years ago during the Late Cambrian or Early Ordovician. Vertebrates such as the lobe-finned fish Tiktaalik started to move on to land in the late Devonian, about 375 million years ago. Animals occupy virtually all of earth's habitats and microhabitats, including salt water, hydrothermal vents, fresh water, hot springs, swamps, forests, pastures, deserts, air, and the interiors of animals, plants, fungi and rocks. Animals are however not particularly heat tolerant; very few of them can survive at constant temperatures above 50 °C (122 °F). Only very few species of animals (mostly nematodes) inhabit the most extreme cold deserts of continental Antarctica.

Predators, such as this ultramarine flycatcher (Ficedula superciliaris), feed on other organisms.

Hydrothermal vent mussels and shrimps.

Diversity

Largest and Smallest

The blue whale (Balaenoptera musculus) is the largest animal that has ever lived, weighing up to 190 metric tonnes and measuring up to 33.6 metres (110 ft) long. The largest extant terrestrial animal is the African bush elephant (Loxodonta africana), weighing up to 12.25 tonnes and measuring up to 10.67 metres (35.0 ft) long. The largest terrestrial animals that ever lived were titanosaur sauropod dinosaurs such as Argentinosaurus, which may have weighed as much as 73 tonnes. Several animals are microscopic; some Myxozoa (obligate parasites within the Cnidaria) never grow larger than 20 µm, and one of the smallest species (Myxobolus shekel) is no more than 8.5 µm when fully grown.

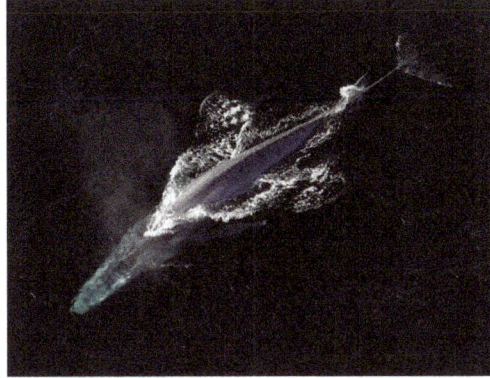

The blue whale is the largest animal that has ever lived.

Evolutionary Origin

The first fossils that might represent animals appear in the 665-million-year-old rocks of the Trezona Formation of South Australia. These fossils are interpreted as most probably being early sponges.

The oldest animals are found in the Ediacaran biota, towards the end of the Precambrian, around 610 million years ago. It had long been doubtful whether these included animals, but the discovery of the animal lipid cholesterol in fossils of Dickinsonia establishes that these were indeed animals. Animals are thought to have originated under low-oxygen conditions, suggesting that they were capable of living entirely by anaerobic respiration, but as they became specialized for aerobic metabolism they became fully dependent on oxygen in their environments.

Many animal phyla first appear in the fossil record during the Cambrian explosion, starting about 542 million years ago, in beds such as the Burgess shale. Extant phyla in these rocks include molluscs, brachiopods, onychophorans, tardigrades, arthropods, echinoderms and hemichordates, along with numerous now-extinct forms such as the predatory Anomalocaris. The apparent suddenness of the event may however be an artefact of the fossil record, rather than showing that all these animals appeared simultaneously.

Dickinsonia costata from the Ediacaran biota (c. 635–542
MYA) is one of the earliest animal species known.

Some palaeontologists have suggested that animals appeared much earlier than the Cambrian explosion, possibly as early as 1 billion years ago. Trace fossils such as tracks and burrows found in the Tonian period may indicate the presence of triploblastic worm-like animals, roughly as large (about 5 mm wide) and complex as earthworms. However, similar tracks are produced today by the giant single-celled protist Gromia sphaerica, so the Tonian trace fossils may not indicate early animal evolution. Around the same time, another line of evidence may indicate the appearance of grazing animals: the layered mats of microorganisms called stromatolites decreased in diversity, perhaps due to grazing.

Anomalocaris canadensis is one of the many animal species that emerged in the Cambrian explosion, starting some 542 million years ago, and found in the fossil beds of the Burgess shale.

Phylogeny

Animals are monophyletic, meaning they are derived from a common ancestor. Animals are sister to the Choanoflagellata, with which they form the Choanozoa. The most basal animals, the Porifera, Ctenophora, Cnidaria, and Placozoa, have body plans that lack bilateral symmetry. Their relationships are still disputed; the sister group to all other animals could be the Porifera or the Ctenophora, which like the Porifera lack hox genes, important in body plan development.

These genes are found in the Placozoa and the higher animals, the Bilateria. 6,331 groups of genes common to all living animals have been identified; these may have arisen from a single common ancestor that lived 650 million years ago in the Precambrian. 25 of these are novel core gene groups, found only in animals; of those, 8 are for essential components of the Wnt and TGF-beta signalling pathways which may have enabled animals to become multicellular by providing a pattern for the body's system of axes (in three dimensions), and another 7 are for transcription factors including homeodomain proteins involved in the control of development.

The phylogenetic tree (of major lineages only) indicates approximately how many millions of years ago (mya) the lineages split.

Non-Bilaterian Animals

Several animal phyla lack bilateral symmetry. Among these, the sponges (Porifera) probably diverged first, representing the oldest animal phylum. Sponges lack the complex organization found in most other animal phyla; their cells are differentiated, but in most cases not organised into distinct tissues. They typically feed by drawing in water through pores.

The Ctenophora (comb jellies) and Cnidaria (which includes jellyfish, sea anemones, and corals) are radially symmetric and have digestive chambers with a single opening, which serves as both mouth and anus. Animals in both phyla have distinct tissues, but these are not organised into organs. They are diploblastic, having only two main germ layers, ectoderm and endoderm. The tiny placozoans are similar, but they do not have a permanent digestive chamber.

Non-bilaterians include sponges (centre) and corals (background).

Bilaterian Animals

The remaining animals, the great majority—comprising some 29 phyla and over a million species—form a clade, the Bilateria. The body is triploblastic, with three well-developed germ layers, and their tissues form distinct organs. The digestive chamber has two openings, a mouth and an anus, and there is an internal body cavity, a coelom or pseudocoelom. Animals with this bilaterally symmetric body plan and a tendency to move in one direction have a head end (anterior) and a tail end (posterior) as well as a back (dorsal) and a belly (ventral); therefore they also have a left side and a right side.

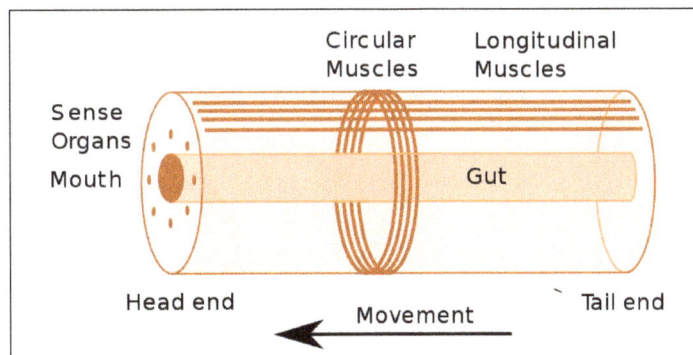

Idealised bilaterian body plan.With an elongated body and a direction of movement the animal has head and tail ends. Sense organs and mouth form the basis of the head. Opposed circular and longitudinal muscles enable peristaltic motion.

Having a front end means that this part of the body encounters stimuli, such as food, favouring cephalisation, the development of a head with sense organs and a mouth. Many bilaterians have a combination of circular muscles that constrict the body, making it longer, and an opposing set of longitudinal muscles, that shorten the body; these enable soft-bodied animals with a hydrostatic skeleton to move by peristalsis. They also have a gut that extends through the basically cylindrical body from mouth to anus. Many bilaterian phyla have primary larvae which swim with cilia and have an apical organ containing sensory cells. However, there are exceptions to each of these characteristics; for example, adult echinoderms are radially symmetric (unlike their larvae), while some parasitic worms have extremely simplified body structures.

Genetic studies have considerably changed zoologists' understanding of the relationships within the Bilateria. Most appear to belong to two major lineages, the protostomes and the deuterostomes. The basalmost bilaterians are the Xenacoelomorpha.

Protostomes and Deuterostomes

Protostomes and deuterostomes differ in several ways. Early in development, deuterostome embryos undergo radial cleavage during cell division, while many protostomes (the Spiralia) undergo spiral cleavage. Animals from both groups possess a complete digestive tract, but in protostomes the first opening of the embryonic gut develops into the mouth, and the anus forms secondarily. In deuterostomes, the anus forms first while the mouth develops secondarily. Most protostomes have schizocoelous development, where cells simply fill in the interior of the gastrula to form the mesoderm. In deuterostomes, the mesoderm forms by enterocoelic pouching, through invagination of the endoderm.

The main deuterostome phyla are the Echinodermata and the Chordata. Echinoderms are exclusively marine and include starfish, sea urchins, and sea cucumbers. The chordates are dominated by the vertebrates (animals with backbones), which consist of fishes, amphibians, reptiles, birds, and mammals. The deuterostomes also include the Hemichordata (acorn worms).

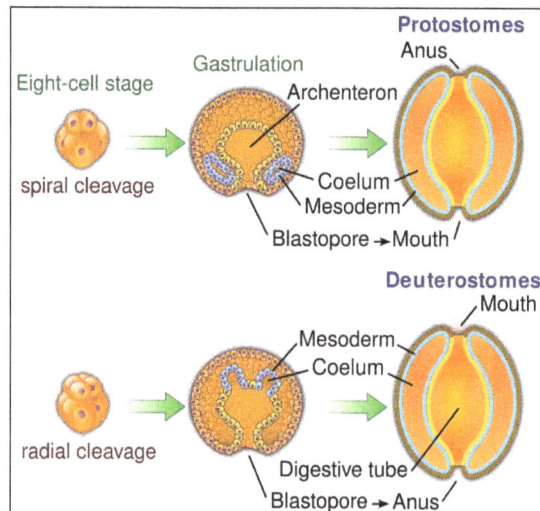

The bilaterian gut develops in two ways. In many protostomes, the blastopore develops into the mouth, while in deuterostomes it becomes the anus.

Ecdysozoa

The Ecdysozoa are protostomes, named after their shared trait of ecdysis, growth by moulting. They include the largest animal phylum, the Arthropoda, which contains insects, spiders, crabs, and their kin. All of these have a body divided into repeating segments, typically with paired appendages. Two smaller phyla, the Onychophora and Tardigrada, are close relatives of the arthropods and share these traits. The ecdysozoans also include the Nematoda or roundworms, perhaps the second largest animal phylum. Roundworms are typically microscopic, and occur in nearly every environment where there is water; some are important parasites. Smaller phyla related to them are the Nematomorpha or horsehair worms, and the Kinorhyncha, Priapulida, and Loricifera. These groups have a reduced coelom, called a pseudocoelom.

Ecdysis: a dragonfly has emerged from its dry exuviae and is expanding
its wings. Like other arthropods,its body is divided into segments.

Spiralia

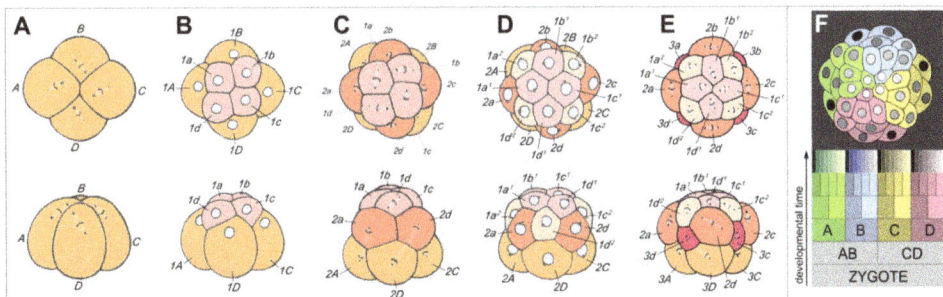

Spiral cleavage in a sea snail embryo.

The Spiralia are a large group of protostomes that develop by spiral cleavage in the early embryo. The Spiralia's phylogeny has been disputed, but it contains a large clade, the superphylum Lophotrochozoa, and smaller groups of phyla such as the Rouphozoa which includes the gastrotrichs and the flatworms. All of these are grouped as the Platytrochozoa, which has a sister group, the Gnathifera, which includes the rotifers.

The Lophotrochozoa includes the molluscs, annelids, brachiopods, nemerteans, bryozoa and entoprocts. The molluscs, the second-largest animal phylum by number of described species, includes snails, clams, and squids, while the annelids are the segmented worms, such as earthworms, lugworms, and leeches. These two groups have long been considered close relatives because they share trochophore larvae.

In Human Culture

The human population exploits a large number of other animal species for food, both of domesticated livestock species in animal husbandry and, mainly at sea, by hunting wild species. Marine fish of many species are caught commercially for food. A smaller number of species are farmed commercially. Invertebrates including cephalopods, crustaceans, and bivalve or gastropod molluscs are hunted or farmed for food. Chickens, cattle, sheep, pigs and other animals are raised as livestock for meat across the world. Animal fibres such as wool are used to make textiles, while animal sinews have been used as lashings and bindings, and leather is widely used to make shoes and other items. Animals have been hunted and farmed for their fur to make items such as coats and hats. Dyestuffs including carmine (cochineal), shellac, and kermes have been made from the bodies of insects. Working animals including cattle and horses have been used for work and transport from the first days of agriculture.

Animals such as the fruit fly Drosophila melanogaster serve a major role in science as experimental models. Animals have been used to create vaccines since their discovery in the 18th century. Some medicines such as the cancer drug Yondelis are based on toxins or other molecules of animal origin.

People have used hunting dogs to help chase down and retrieve animals, and birds of prey to catch birds and mammals, while tethered cormorants have been used to catch fish. Poison dart frogs have been used to poison the tips of blowpipe darts. A wide variety of animals are kept as pets, from invertebrates such as tarantulas and octopuses, insects including praying mantises, reptiles such as snakes and chameleons, and birds including canaries, parakeets, and parrots all finding a place. However, the most kept pet species are mammals, namely dogs, cats, and rabbits. There is a tension between the role of animals as companions to humans, and their existence as individuals with rights of their own. A wide variety of terrestrial and aquatic animals are hunted for sport.

Sides of beef in a slaughterhouse.

Animals have been the subjects of art from the earliest times, both historical, as in Ancient Egypt, and prehistoric, as in the cave paintings at Lascaux. Major animal paintings include Albrecht Dürer's 1515 The Rhinoceros, and George Stubbs's c. 1762 horse portrait Whistlejacket. Insects, birds and mammals play roles in literature and film, such as in giant bug movies. Animals including insects and mammals feature in mythology and religion. In both Japan and Europe, a butterfly was seen as the personification of a person's soul, while the scarab beetle was sacred in ancient

Egypt. Among the mammals, cattle, deer, horses, lions, bats, bears, and wolves are the subjects of myths and worship. The signs of the Western and Chinese zodiacs are based on animals.

A gun dog retrieving a duck during a hunt.

Still Life with Lobster and Oysters.

DOMAIN

In biological taxonomy, a domain also superkingdom or empire, is the highest taxonomic rank of organisms in the three-domain system of taxonomy designed by Carl Woese et.al. in 1990.

According to this system, the tree of life consists of three domains: Archaea, Bacteria, and Eukarya. The first two are all prokaryotic microorganisms, or single-celled organisms whose cells have no nucleus. All life that has a nucleus and membrane-bound organelles, and multicellular organisms, is included in the Eukarya.

Dominion

The term "domain" was proposed by Carl Woese, Otto Kandler and Mark Wheelis (1990) in a three-domain system. This term represents a synonym for the category of dominion (Lat. dominium), introduced by Moore in 1974. However, only Stefan Luketa uses the term "dominion". He created two additional domains ("dominions") for Prions and Viruses.

Characteristics of the Three Domains

Each of these three domains contains unique rRNA. This forms the basis of the three-domain system. While the presence of a nuclear membrane differentiates the Eukarya from the Archaea and Bacteria, both of which lack a nuclear membrane, distinct biochemical and RNA markers differentiate the Archaea and Bacteria from each other.

Archaea

Archaea are prokaryota cells, typically characterized by membrane lipids that are branched hydrocarbon chains attached to glycerol by ether linkages. The presence of these other linkages in Archaea adds to their ability to withstand extreme temperatures and highly acidic conditions, but many archeae live in mild environments. Halophiles, organisms that thrive in highly salty environments, and hyperthermophiles, organisms that thrive in extremely hot environments, are examples of Archaea.

Archaea evolved many cell sizes, but all are relatively small. Their size ranges from 0.1 μm to 15 μm diameter and up to 200 μm long. They are about the size of bacteria, or similar in size to the mitochondria found in eukaryotic cells. Members of the genus Thermoplasma are the smallest of the Archaea.

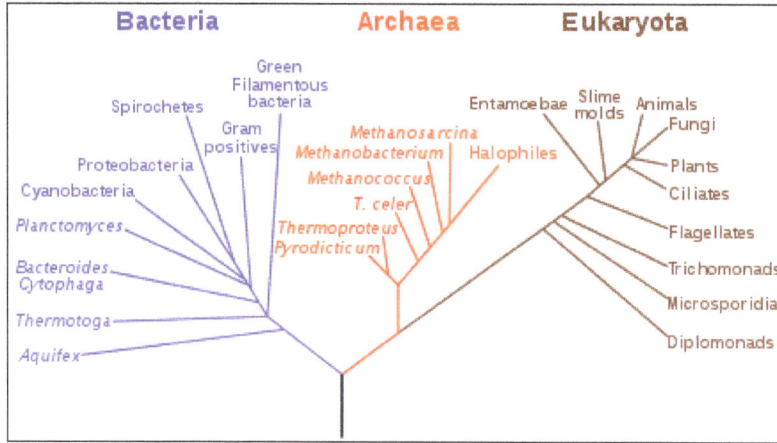

A speculatively rooted tree for RNA genes, showing major branches Bacteria, Archaea, and Eukaryota.

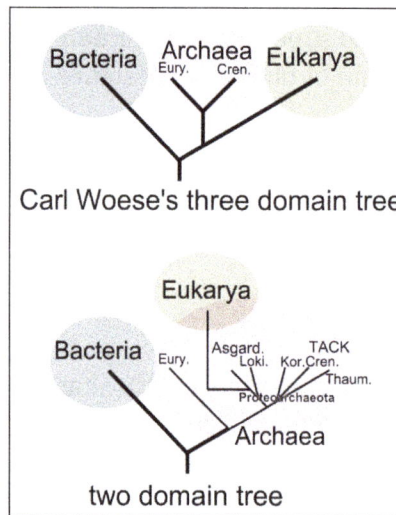

The three-domains tree and the Eocyte hypothesis (Two domains tree).

Bacteria

Even though bacteria are prokaryotic cells just like Archaea, their membranes are made of oligonucleotide acid attached to glycerol by ester linkages. Cyanobacteria and mycoplasmas are two examples of bacteria. They characteristically do not have ether linkages like Archaea, and they are grouped into a different category—and hence a different domain. There is a great deal of diversity in this domain. Confounded by that diversity and horizontal gene transfer, it is next to impossible to determine how many species of bacteria exist on the planet, or to organize them in a tree-structure, without cross-connections between branches.

Eukarya

Members of the domain Eukarya—called eukaryotes—have membrane-bound organelles (including a nucleus containing genetic material) and are represented by five kingdoms: Plantae, Protista, Animalia, Chromista, and Fungi.

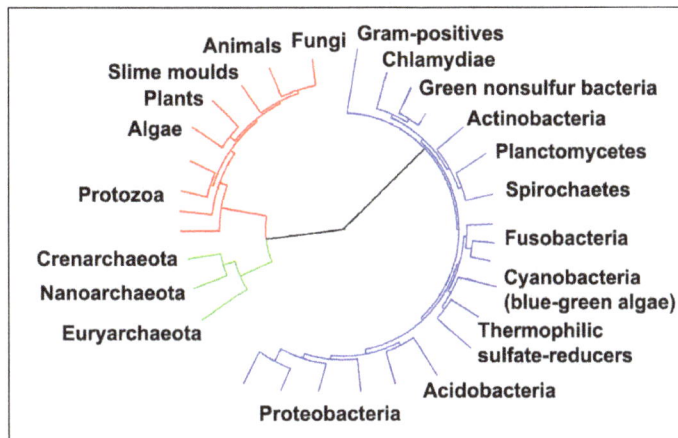

Phylogenetic tree showing the relationship between the eukaryotes and other forms of life, 2006 Eukaryotes are colored red, archaea green, and bacteria blue.

Exclusion of Viruses and Prions

The three-domain system does not include any form of non-cellular life. As of 2011 there was talk about nucleocytoplasmic large DNA viruses possibly being a fourth domain of life, a view supported by researchers in 2012. Stefan Luketa proposed a five-domain system in 2012, adding Prionobiota (acellular and without nucleic acid) and Virobiota (acellular but with nucleic acid) to the traditional three domains.

Alternative Classifications

Alternative classifications of life include:

- The two-empire system or superdomain system, with top-level groupings of Prokaryota (or Monera) and Eukaryota.

- The eocyte hypothesis, first proposed by James A. Lake et al. in 1984, which posits two domains (Bacteria and Archaea, with Eukaryota included in Archaea).

Archaea

Archaea (singular archaeon) constitute a domain of single-celled organisms. These microorganisms are prokaryotes, and have no cell nucleus. Archaea were initially classified as bacteria, receiving the name archaebacteria (in the Archaebacteria kingdom), but this classification is outmoded.

Archaeal cells have unique properties separating them from the other two domains of Bacteria and Eukaryota. Archaea are further divided into multiple recognized phyla. Classification is difficult because most have not been isolated in the laboratory and have only been detected by analysis of their nucleic acids in samples from their environment.

Archaea and bacteria are generally similar in size and shape, although a few archaea have very different shapes, such as the flat and square cells of Haloquadratum walsbyi. Despite this morphological similarity to bacteria, archaea possess genes and several metabolic pathways that are more closely related to those of eukaryotes, notably for the enzymes involved in transcription and translation. Other aspects of archaeal biochemistry are unique, such as their reliance on ether lipids in their cell membranes, including archaeols. Archaea use more energy sources than eukaryotes: these range from organic compounds, such as sugars, to ammonia, metal ions or even hydrogen gas. Salt-tolerant archaea (the Haloarchaea) use sunlight as an energy source, and other species of archaea fix carbon, but unlike plants and cyanobacteria, no known species of archaea does both. Archaea reproduce asexually by binary fission, fragmentation, or budding; unlike bacteria and eukaryotes, no known species forms spores.

The first observed archaea were extremophiles, living in harsh environments, such as hot springs and salt lakes with no other organisms, but improved detection tools led to the discovery of archaea in almost every habitat, including soil, oceans, and marshlands. They are also part of the microbiota of all organisms, and in the human microbiota they are important in the gut, mouth, and on the skin. Archaea are particularly numerous in the oceans, and the archaea in plankton may be one of the most abundant groups of organisms on the planet. Archaea are a major part of Earth's life, and may play roles in the carbon cycle and the nitrogen cycle. No clear examples of archaeal pathogens or parasites are known. Instead they are often mutualists or commensals, such as the methanogens (methane-producing strains) that inhabit the gastrointestinal tract in humans and ruminants, where their vast numbers aid digestion. Methanogens are also used in biogas production and sewage treatment, and biotechnology exploits enzymes from extremophile archaea that can endure high temperatures and organic solvents.

Classification

Early Concept

For much of the 20th century, prokaryotes were regarded as a single group of organisms and classified based on their biochemistry, morphology and metabolism. Microbiologists tried to classify microorganisms based on the structures of their cell walls, their shapes, and the substances they consume. In 1965, Emile Zuckerkandl and Linus Pauling proposed instead using the sequences of the genes in different prokaryotes to work out how they are related to each other. This phylogenetic approach is the main method used today.

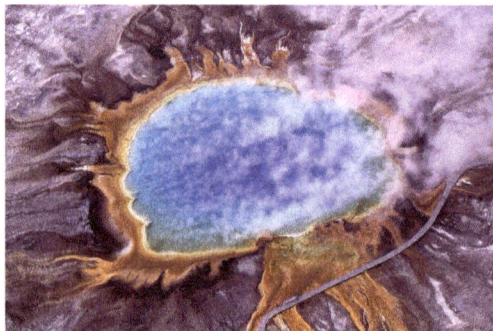

Archaea were found in volcanic hot springs. Pictured here is
Grand Prismatic Springof Yellowstone National Park.

Archaea — at that time only the methanogens were known — were first classified separately from bacteria in 1977 by Carl Woese and George E. Fox based on their ribosomal RNA (rRNA) genes.

They called these groups the Urkingdoms of Archaebacteria and Eubacteria, though other researchers treated them as kingdoms or subkingdoms. Woese and Fox gave the first evidence for Archaebacteria as a separate "line of descent": 1. lack of peptidoglycan in their cell walls, 2. two unusual coenzymes, 3. results of 16S ribosomal RNA gene sequencing. To emphasize this difference, Woese, Otto Kandler and Mark Wheelis later proposed reclassifying organisms into three natural domains known as the three-domain system: the Eukarya, the Bacteria and the Archaea, in what is now known as "The Woesian Revolution".

The first representatives of the domain Archaea were methanogens and it was assumed that their metabolism reflected Earth's primitive atmosphere and the organisms' antiquity, but as new habitats were studied, more organisms were discovered. Extreme halophilic and hyperthermophilic microbes were also included in Archaea. For a long time, archaea were seen as extremophiles that only exist in extreme habitats such as hot springs and salt lakes, but by the end of the 20th century, archaea had been identified in non-extreme environments as well. Today, they are known to be a large and diverse group of organisms abundantly distributed throughout nature. This new appreciation of the importance and ubiquity of archaea came from using polymerase chain reaction (PCR) to detect prokaryotes from environmental samples (such as water or soil) by multiplying their ribosomal genes. This allows the detection and identification of organisms that have not been cultured in the laboratory.

Current Classification

The ARMAN are a new group of archaea
recently discovered in acid mine drainage.

The classification of archaea, and of prokaryotes in general, is a rapidly moving and contentious field. Current classification systems aim to organize archaea into groups of organisms that share structural features and common ancestors. These classifications rely heavily on the use of the sequence of ribosomal RNA genes to reveal relationships between organisms (molecular phylogenetics). Most of the culturable and well-investigated species of archaea are members of two main phyla, the Euryarchaeota and Crenarchaeota. Other groups have been tentatively created, like the peculiar species Nanoarchaeum equitans, which was discovered in 2003, has been given its own phylum, the Nanoarchaeota. A new phylum Korarchaeota has also been proposed. It contains a small group of unusual thermophilic species that shares features of both of the main phyla, but is most closely related to the Crenarchaeota. Other recently detected species of archaea are only distantly related to any of these groups, such as the Archaeal Richmond Mine acidophilic nanoorganisms (ARMAN, comprising Micrarchaeota and Parvarchaeota), which were discovered in 2006 and are some of the smallest organisms known.

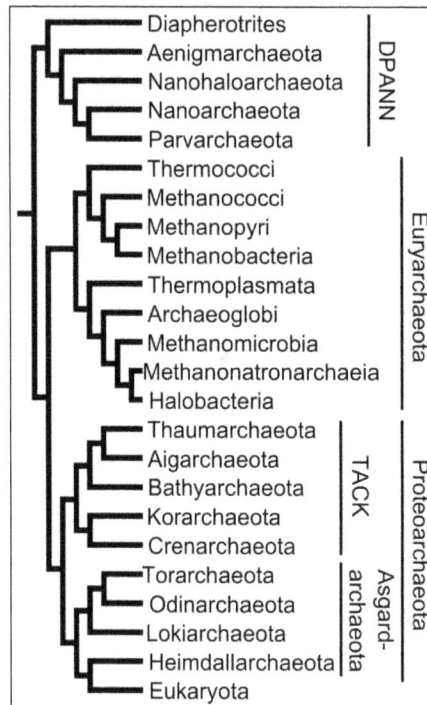

Phylogenetic tree of Archaea using conserved genes.

A superphylum – TACK – has been proposed that includes the Thaumarchaeota, Aigarchaeota, Crenarchaeota, and Korarchaeota. This superphylum may be related to the origin of eukaryotes. More recently, the superphylum Asgard has been named and proposed to be more closely related to the original eukaryote and a sister group to TACK.

Concept of Species

The classification of archaea into species is also controversial. Biology defines a species as a group of related organisms. The familiar exclusive breeding criterion (organisms that can breed with each other but not with others) is of no help since archaea reproduce asexually.

Archaea show high levels of horizontal gene transfer between lineages. Some researchers suggest that individuals can be grouped into species-like populations given highly similar genomes and infrequent gene transfer to/from cells with less-related genomes, as in the genus Ferroplasma. On the other hand, studies in Halorubrum found significant genetic transfer to/from less-related populations, limiting the criterion's applicability. Some researchers question whether such species designations have practical meaning.

Current knowledge on genetic diversity is fragmentary and the total number of archaeal species cannot be estimated with any accuracy. Estimates of the number of phyla range from 18 to 23, of which only 8 have representatives that have been cultured and studied directly. Many of these hypothesized groups are known from a single rRNA sequence, indicating that the diversity among these organisms remains obscure. The Bacteria also include many uncultured microbes with similar implications for characterization.

On average, archaeal genomes show higher levels of complexity than those of bacteria.

Origin and Evolution

The age of the Earth is about 4.54 billion years. Scientific evidence suggests that life began on Earth at least 3.5 billion years ago. The earliest evidence for life on Earth is graphite found to be biogenic in 3.7 billion-year-old metasedimentary rocks discovered in Western Greenland and microbial mat fossils found in 3.48 billion-year-old sandstone discovered in Western Australia. In 2015, possible remains of biotic matter were found in 4.1 billion-year-old rocks in Western Australia.

Although probable prokaryotic cell fossils date to almost 3.5 billion years ago, most prokaryotes do not have distinctive morphologies and fossil shapes cannot be used to identify them as archaea. Instead, chemical fossils of unique lipids are more informative because such compounds do not occur in other organisms. Some publications suggest that archaeal or eukaryotic lipid remains are present in shales dating from 2.7 billion years ago; such data have since been questioned. Such lipids have also been detected in even older rocks from west Greenland. The oldest such traces come from the Isua district, which includes Earth's oldest known sediments, formed 3.8 billion years ago. The archaeal lineage may be the most ancient that exists on Earth.

Woese argued that the Bacteria, Archaea, and Eukaryotes represent separate lines of descent that diverged early on from an ancestral colony of organisms. One possibility is that this occurred before the evolution of cells, when the lack of a typical cell membrane allowed unrestricted lateral gene transfer, and that the common ancestors of the three domains arose by fixation of specific subsets of genes. It is possible that the last common ancestor of bacteria and archaea was a thermophile, which raises the possibility that lower temperatures are "extreme environments" for archaea, and organisms that live in cooler environments appeared only later. Since archaea and bacteria are no more related to each other than they are to eukaryotes, the term prokaryote suggests a false similarity between them.

Comparison to other Domains

The following table compares some major characteristics of the three domains, to illustrate their similarities and differences.

Property	Archaea	Bacteria	Eukarya
Cell membrane	Ether-linked lipids	Ester-linked lipids	Ester-linked lipids
Cell wall	Pseudopeptidoglycan, glycoprotein, or S-layer	Peptidoglycan, S-layer, or no cell wall	Various structures
Gene structure	Circular chromosomes, similar translation and transcription to Eukarya	Circular chromosomes, unique translation and transcription	Multiple, linear chromosomes, but translation and transcription similar to Archaea
Internal cell structure	No membrane-bound organelles (?) or nucleus	No membrane-bound organelles or nucleus	Membrane-bound organelles and nucleus
Metabolism	Various, including diazotrophy, with methanogenesis unique to Archaea	Various, including photosynthesis, aerobic and anaerobic respiration, fermentation, diazotrophy, and autotrophy	Photosynthesis, cellular respiration, and fermentation; no diazotrophy
Reproduction	Asexual reproduction, horizontal gene transfer	Asexual reproduction, horizontal gene transfer	Sexual and asexual reproduction

Archaea were split off as a third domain because of the large differences in their ribosomal RNA structure. The particular molecule 16S rRNA is key to the production of proteins in all organisms. Because this function is so central to life, organisms with mutations in their 16S rRNA are unlikely to survive, leading to great (but not absolute) stability in the structure of this nucleotide over generations. 16S rRNA is large enough to show organism-specific variations, but still small enough to be compared quickly. In 1977, Carl Woese, a microbiologist studying the genetic sequences of organisms, developed a new comparison method that involved splitting the RNA into fragments that could be sorted and compared to other fragments from other organisms. The more similar the patterns between species, the more closely they are related.

Woese used his new rRNA comparison method to categorize and contrast different organisms. He compared a variety of species and happened upon a group of methanogens with rRNA vastly different from any known prokaryotes or eukaryotes. These methanogens were much more similar to each other than to other organisms, leading Woese to propose the new domain of Archaea. His experiments showed that the archaea were genetically more similar to eukaryotes than prokaryotes, even though they were more similar to prokaryotes in structure. This led to the conclusion that Archaea and Eukarya shared a more recent common ancestor than Eukarya and Bacteria. The development of the nucleus occurred after the split between Bacteria and this common ancestor.

One property unique to archaea is the abundant use of ether-linked lipids in their cell membranes. Ether linkages are more chemically stable than the ester linkages found in bacteria and eukarya, which may be a contributing factor to the ability of many archaea to survive in extreme environments that place heavy stress on cell membranes, such as extreme heat and salinity. Comparative analysis of archaeal genomes has also identified several molecular conserved signature indels and signature proteins uniquely present in either all archaea or different main groups within archaea. Another unique feature of archaea, found in no other organisms, is methanogenesis (the metabolic production of methane). Methanogenic archaea play a pivotal role in ecosystems with organisms that derive energy from oxidation of methane, many of which are bacteria, as they are often a major source of methane in such environments and can play a role as primary producers. Methanogens also play a critical role in the carbon cycle, breaking down organic carbon into methane, which is also a major greenhouse gas.

Relationship to Bacteria

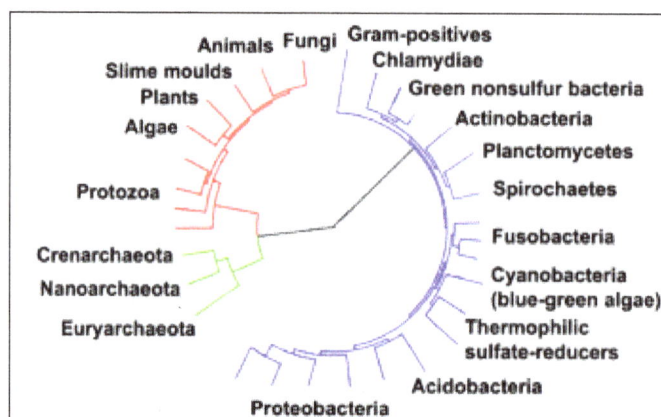

Phylogenetic tree showing the relationship between the Archaea and other domains of life.Eukaryotes are colored red, archaea green and bacteria blue.

The relationship between the three domains is of central importance for understanding the origin of life. Most of the metabolic pathways, which are the object of the majority of an organism's genes, are common between Archaea and Bacteria, while most genes involved in genome expression are common between Archaea and Eukarya. Within prokaryotes, archaeal cell structure is most similar to that of gram-positive bacteria, largely because both have a single lipid bilayer and usually contain a thick sacculus (exoskeleton) of varying chemical composition. In some phylogenetic trees based upon different gene/protein sequences of prokaryotic homologs, the archaeal homologs are more closely related to those of gram-positive bacteria. Archaea and gram-positive bacteria also share conserved indels in a number of important proteins, such as Hsp70 and glutamine synthetase I;, but the phylogeny of these genes was interpreted to reveal interdomain gene transfer, and might not reflect the organismal relationship(s).

It has been proposed that the archaea evolved from gram-positive bacteria in response to antibiotic selection pressure. This is suggested by the observation that archaea are resistant to a wide variety of antibiotics that are primarily produced by gram-positive bacteria, and that these antibiotics primarily act on the genes that distinguish archaea from bacteria. The proposal is that the selective pressure towards resistance generated by the gram-positive antibiotics was eventually sufficient to cause extensive changes in many of the antibiotics' target genes, and that these strains represented the common ancestors of present-day Archaea. The evolution of Archaea in response to antibiotic selection, or any other competitive selective pressure, could also explain their adaptation to extreme environments (such as high temperature or acidity) as the result of a search for unoccupied niches to escape from antibiotic-producing organisms; Cavalier-Smith has made a similar suggestion. This proposal is also supported by other work investigating protein structural relationships and studies that suggest that gram-positive bacteria may constitute the earliest branching lineages within the prokaryotes.

Relation to Eukaryotes

The evolutionary relationship between archaea and eukaryotes remains unclear. Aside from the similarities in cell structure and function that are discussed below, many genetic trees group the two.

Complicating factors include claims that the relationship between eukaryotes and the archaeal phylum Crenarchaeota is closer than the relationship between the Euryarchaeota and the phylum Crenarchaeota and the presence of archaea-like genes in certain bacteria, such as Thermotoga maritima, from horizontal gene transfer. The standard hypothesis states that the ancestor of the eukaryotes diverged early from the Archaea, and that eukaryotes arose through fusion of an archaean and eubacterium, which became the nucleus and cytoplasm; this hypothesis explains various genetic similarities but runs into difficulties explaining cell structure. An alternative hypothesis, the eocyte hypothesis, posits that Eukaryota emerged relatively late from the Archaea.

A lineage of archaea discovered in 2015, Lokiarchaeum (of proposed new Phylum "Lokiarchaeota"), named for a hydrothermal vent called Loki's Castle in the Arctic Ocean, was found to be the most closely related to eukaryotes known at that time. It has been called a transitional organism between prokaryotes and eukaryotes.

Several sister phyla of "Lokiarchaeota" have since been found ("Thorarchaeota", "Odinarchaeota", "Heimdallarchaeota"), all together comprising a newly proposed supergroup Asgard, which may appear as a sister taxon to Proteoarchaeota. Details of the relation of Asgard members and eukaryotes are still under consideration.

Morphology

Individual archaea range from 0.1 micrometers (μm) to over 15 μm in diameter, and occur in various shapes, commonly as spheres, rods, spirals or plates. Other morphologies in the Crenarchaeota include irregularly shaped lobed cells in Sulfolobus, needle-like filaments that are less than half a micrometer in diameter in Thermofilum, and almost perfectly rectangular rods in Thermoproteus and Pyrobaculum. Archaea in the genus Haloquadratum such as Haloquadratum walsbyi are flat, square archaea that live in hypersaline pools. These unusual shapes are probably maintained both by their cell walls and a prokaryotic cytoskeleton. Proteins related to the cytoskeleton components of other organisms exist in archaea, and filaments form within their cells, but in contrast to other organisms, these cellular structures are poorly understood. In Thermoplasma and Ferroplasma the lack of a cell wall means that the cells have irregular shapes, and can resemble amoebae.

Some species form aggregates or filaments of cells up to 200 μm long. These organisms can be prominent in biofilms. Notably, aggregates of Thermococcus coalescens cells fuse together in culture, forming single giant cells. Archaea in the genus Pyrodictium produce an elaborate multicell colony involving arrays of long, thin hollow tubes called cannulae that stick out from the cells' surfaces and connect them into a dense bush-like agglomeration. The function of these cannulae is not settled, but they may allow communication or nutrient exchange with neighbors. Multi-species colonies exist, such as the "string-of-pearls" community that was discovered in 2001 in a German swamp. Round whitish colonies of a novel Euryarchaeota species are spaced along thin filaments that can range up to 15 centimetres (5.9 in) long; these filaments are made of a particular bacteria species.

Structure, Composition Development and Operation

Archaea and bacteria have generally similar cell structure, but cell composition and organization set the archaea apart. Like bacteria, archaea lack interior membranes and organelles. Like bacteria, the cell membranes of archaea are usually bounded by a cell wall and they swim using one or more flagella. Structurally, archaea are most similar to gram-positive bacteria. Most have a single plasma membrane and cell wall, and lack a periplasmic space; the exception to this general rule is Ignicoccus, which possess a particularly large periplasm that contains membrane-bound vesicles and is enclosed by an outer membrane.

Cell Wall and Flagella

Most archaea (but not Thermoplasma and Ferroplasma) possess a cell wall. In most archaea the wall is assembled from surface-layer proteins, which form an S-layer. An S-layer is a rigid array of protein molecules that cover the outside of the cell (like chain mail). This layer provides both chemical and physical protection, and can prevent macromolecules from contacting the cell membrane. Unlike bacteria, archaea lack peptidoglycan in their cell walls. Methanobacteriales do have cell walls containing pseudopeptidoglycan, which resembles eubacterial peptidoglycan in morphology, function, and physical structure, but pseudopeptidoglycan is distinct in chemical structure; it lacks D-amino acids and N-acetylmuramic acid.

Archaea flagella are known as archaella, that operate like bacterial flagella – their long stalks are driven by rotatory motors at the base. These motors are powered by the proton gradient across the membrane, but archaealla are notably different in composition and development. The two types of flagella evolved from different ancestors. The bacterial flagellum shares a common ancestor with the type III secretion system, while archaeal flagella appear to have evolved from bacterial type IV pili. In contrast to the bacterial flagellum, which is hollow and is assembled by subunits moving up the central pore to the tip of the flagella, archaeal flagella are synthesized by adding subunits at the base.

Membranes

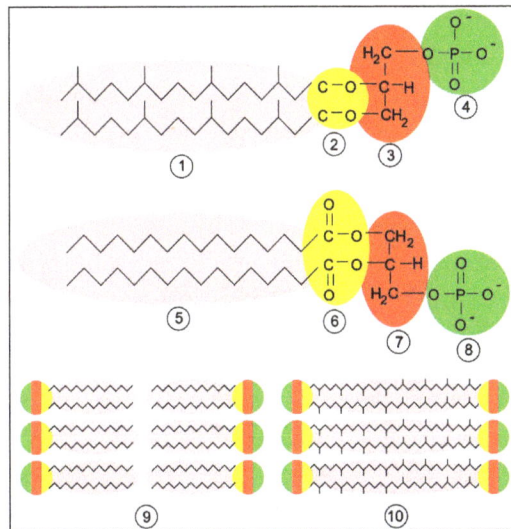

In figure, membrane structures. Top, an archaeal phospholipid: 1, isoprene chains; 2, ether linkages; 3, L-glycerol moiety; 4, phosphate group. Middle, a bacterial or eukaryotic phospholipid: 5, fatty acid chains; 6, ester linkages; 7, D-glycerol moiety; 8, phosphate group. Bottom: 9, lipid bilayer of bacteria and eukaryotes; 10, lipid monolayer of some archaea.

Archaeal membranes are made of molecules that are distinctly different from those in all other life forms, showing that archaea are related only distantly to bacteria and eukaryotes. In all organisms, cell membranes are made of molecules known as phospholipids. These molecules possess both a polar part that dissolves in water (the phosphate "head"), and a "greasy" non-polar part that does not (the lipid tail). These dissimilar parts are connected by a glycerol moiety. In water, phospholipids cluster, with the heads facing the water and the tails facing away from it. The major structure in cell membranes is a double layer of these phospholipids, which is called a lipid bilayer.

The phospholipids of archaea are unusual in four ways:

- They have membranes composed of glycerol-ether lipids, whereas bacteria and eukaryotes have membranes composed mainly of glycerol-ester lipids. The difference is the type of bond that joins the lipids to the glycerol moiety; the two types are shown in yellow in the figure at the right. In ester lipids this is an ester bond, whereas in ether lipids this is an ether bond. Ether bonds are chemically more resistant than ester bonds.

- The stereochemistry of the archaeal glycerol moiety is the mirror image of that found in other organisms. The glycerol moiety can occur in two forms that are mirror images of one another, called enantiomers. Just as a right hand does not fit easily into a left-handed glove, enantiomers of one type generally cannot be used or made by enzymes adapted for the other. The archaeal phospholipids are built on a backbone of sn-glycerol-1-phosphate, which is an enantiomer of sn-glycerol-3-phosphate, the phospholipid backbone found in bacteria and eucaryotes. This suggests that archaea use entirely different enzymes for synthesizing phospholipids than do bacteria and eukaryotes. Such enzymes developed very early in life's history, indicating an early split from the other two domains.

- Archaeal lipid tails differ from those of other organisms in that they are based upon long isoprenoid chains with multiple side-branches, sometimes with cyclopropane or cyclohexane rings. By contrast, the fatty acids in the membranes of other organisms have straight chains without side branches or rings. Although isoprenoids play an important role in the biochemistry of many organisms, only the archaea use them to make phospholipids. These branched chains may help prevent archaeal membranes from leaking at high temperatures.

- In some archaea, the lipid bilayer is replaced by a monolayer. In effect, the archaea fuse the tails of two phospholipid molecules into a single molecule with two polar heads (a bolaamphiphile); this fusion may make their membranes more rigid and better able to resist harsh environments. For example, the lipids in Ferroplasma are of this type, which is thought to aid this organism's survival in its highly acidic habitat.

Metabolism

Archaea exhibit a great variety of chemical reactions in their metabolism and use many sources of energy. These reactions are classified into nutritional groups, depending on energy and carbon sources. Some archaea obtain energy from inorganic compounds such as sulfur or ammonia (they are chemotrophs). These include nitrifiers, methanogens and anaerobic methane oxidisers. In these reactions one compound passes electrons to another (in a redox reaction), releasing energy to fuel the cell's activities. One compound acts as an electron donor and one as an electron acceptor. The energy released is used to generate adenosine triphosphate (ATP) through chemiosmosis, the same basic process that happens in the mitochondrion of eukaryotic cells.

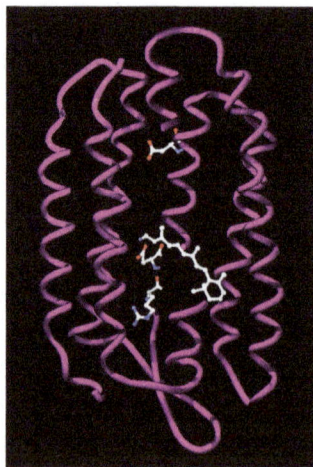

Bacteriorhodopsin from Halobacterium salinarum. The retinol cofactor
and residues involved in proton transfer are shown as ball-and-stick models.

Other groups of archaea use sunlight as a source of energy (they are phototrophs), but oxygen–generating photosynthesis does not occur in any of these organisms. Many basic metabolic pathways are shared between all forms of life; for example, archaea use a modified form of glycolysis (the Entner–Doudoroff pathway) and either a complete or partial citric acid cycle. These similarities to other organisms probably reflect both early origins in the history of life and their high level of efficiency.

Table: Nutritional types in archaeal metabolism.

Nutritional type	Source of energy	Source of carbon	Examples
Phototrophs	Sunlight	Organic compounds	Halobacterium
Lithotrophs	Inorganic compounds	Organic compounds or carbon fixation	Ferroglobus, Methanobacteria or Pyrolobus
Organotrophs	Organic compounds	Organic compounds or carbon fixation	Pyrococcus, Sulfolobus or Methanosarcinales

Some Euryarchaeota are methanogens (archaea that produce methane as a result of metabolism) living in anaerobic environments, such as swamps. This form of metabolism evolved early, and it is even possible that the first free-living organism was a methanogen. A common reaction involves the use of carbon dioxide as an electron acceptor to oxidize hydrogen. Methanogenesis involves a range of coenzymes that are unique to these archaea, such as coenzyme M and methanofuran. Other organic compounds such as alcohols, acetic acid or formic acid are used as alternative electron acceptors by methanogens. These reactions are common in gut-dwelling archaea. Acetic acid is also broken down into methane and carbon dioxide directly, by acetotrophic archaea. These acetotrophs are archaea in the order Methanosarcinales, and are a major part of the communities of microorganisms that produce biogas.

Other archaea use CO_2 in the atmosphere as a source of carbon, in a process called carbon fixation (they are autotrophs). This process involves either a highly modified form of the Calvin cycle or another metabolic pathway called the 3-hydroxypropionate/4-hydroxybutyrate cycle. The Crenarchaeota also use the reverse Krebs cycle while the Euryarchaeota also use the reductive acetyl-CoA pathway. Carbon–fixation is powered by inorganic energy sources. No known archaea carry out photosynthesis. Archaeal energy sources are extremely diverse, and range from the oxidation of ammonia by the Nitrosopumilales to the oxidation of hydrogen sulfide or elemental sulfur by species of Sulfolobus, using either oxygen or metal ions as electron acceptors.

Phototrophic archaea use light to produce chemical energy in the form of ATP. In the Halobacteria, light-activated ion pumps like bacteriorhodopsin and halorhodopsin generate ion gradients by pumping ions out of and into the cell across the plasma membrane. The energy stored in these electrochemical gradients is then converted into ATP by ATP synthase. This process is a form of photophosphorylation. The ability of these light-driven pumps to move ions across membranes depends on light-driven changes in the structure of a retinol cofactor buried in the center of the protein.

Genetics

Archaea usually have a single circular chromosome, with as many as 5,751,492 base pairs in Methanosarcina acetivorans, the largest known archaeal genome. The tiny 490,885 base-pair genome of Nanoarchaeum equitans is one-tenth of this size and the smallest archaeal genome known; it is estimated to contain only 537 protein-encoding genes. Smaller independent pieces of DNA, called plasmids, are also found in archaea. Plasmids may be transferred between cells by physical contact, in a process that may be similar to bacterial conjugation.

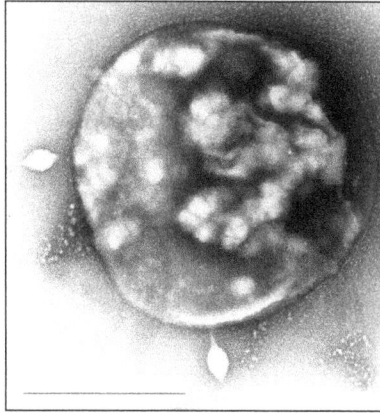

Sulfolobus infected with the DNA virus STSV1. Bar is 1 micrometer.

Archaea can be infected by double-stranded DNA viruses that are unrelated to any other form of virus and have a variety of unusual shapes, including bottles, hooked rods, or teardrops. These viruses have been studied in most detail in thermophilics, particularly the orders Sulfolobales and Thermoproteales. Two groups of single-stranded DNA viruses that infect archaea have been recently isolated. One group is exemplified by the Halorubrum pleomorphic virus 1 ("Pleolipoviridae") infecting halophilic archaea and the other one by the Aeropyrum coil-shaped virus ("Spiraviridae") infecting a hyperthermophilic (optimal growth at 90–95 °C) host. Notably, the latter virus has the largest currently reported ssDNA genome. Defenses against these viruses may involve RNA interference from repetitive DNA sequences that are related to the genes of the viruses.

Archaea are genetically distinct from bacteria and eukaryotes, with up to 15% of the proteins encoded by any one archaeal genome being unique to the domain, although most of these unique genes have no known function. Of the remainder of the unique proteins that have an identified function, most belong to the Euryarchaea and are involved in methanogenesis. The proteins that archaea, bacteria and eukaryotes share form a common core of cell function, relating mostly to transcription, translation, and nucleotide metabolism. Other characteristic archaeal features are the organization of genes of related function – such as enzymes that catalyze steps in the same metabolic pathway into novel operons, and large differences in tRNA genes and their aminoacyl tRNA synthetases.

Transcription in archaea more closely resembles eukaryotic than bacterial transcription, with the archaeal RNA polymerase being very close to its equivalent in eukaryotes; while archaeal translation shows signs of both bacterial and eukaryal equivalents. Although archaea only have one type of RNA polymerase, its structure and function in transcription seems to be close to that of the eukaryotic RNA polymerase II, with similar protein assemblies (the general transcription factors) directing the binding of the RNA polymerase to a gene's promoter, but other archaeal transcription factors are closer to those found in bacteria. Post-transcriptional modification is simpler than in eukaryotes, since most archaeal genes lack introns, although there are many introns in their transfer RNA and ribosomal RNA genes, and introns may occur in a few protein-encoding genes.

Gene Transfer and Genetic Exchange

Halobacterium volcanii, an extreme halophilic archaeon, forms cytoplasmic bridges between cells that appear to be used for transfer of DNA from one cell to another in either direction.

When the hyperthermophilic archaea Sulfolobus solfataricus and Sulfolobus acidocaldarius are exposed to DNA-damaging UV irradiation or to the agents bleomycin or mitomycin C, species-specific cellular aggregation is induced. Aggregation in S. solfataricus could not be induced by other physical stressors, such as pH or temperature shift, suggesting that aggregation is induced specifically by DNA damage. Ajon et al. showed that UV-induced cellular aggregation mediates chromosomal marker exchange with high frequency in S. acidocaldarius. Recombination rates exceeded those of uninduced cultures by up to three orders of magnitude. Frols et al. and Ajon et al. hypothesized that cellular aggregation enhances species-specific DNA transfer between Sulfolobus cells in order to provide increased repair of damaged DNA by means of homologous recombination. This response may be a primitive form of sexual interaction similar to the more well-studied bacterial transformation systems that are also associated with species-specific DNA transfer between cells leading to homologous recombinational repair of DNA damage.

Reproduction

Archaea reproduce asexually by binary or multiple fission, fragmentation, or budding; mitosis and meiosis do not occur, so if a species of archaea exists in more than one form, all have the same genetic material. Cell division is controlled in a cell cycle; after the cell's chromosome is replicated and the two daughter chromosomes separate, the cell divides. In the genus Sulfolobus, the cycle has characteristics that are similar to both bacterial and eukaryotic systems. The chromosomes replicate from multiple starting-points (origins of replication) using DNA polymerases that resemble the equivalent eukaryotic enzymes.

In euryarchaea the cell division protein FtsZ, which forms a contracting ring around the cell, and the components of the septum that is constructed across the center of the cell, are similar to their bacterial equivalents. In cren- and thaumarchaea, but the cell division machinery Cdv fulfills a similar role. This machinery is related to the eukaryotic ESCRT-III machinery which, while best known for its role in cell sorting, also has been seen to fulfill a role in separation between divided cell, suggesting an ancestral role in cell division.

Both bacteria and eukaryotes, but not archaea, make spores. Some species of Haloarchaea undergo phenotypic switching and grow as several different cell types, including thick-walled structures that are resistant to osmotic shock and allow the archaea to survive in water at low salt concentrations, but these are not reproductive structures and may instead help them reach new habitats.

Ecology

Habitats

Archaea exist in a broad range of habitats, and as a major part of global ecosystems, may represent about 20% of microbial cells in the oceans. The first-discovered archaeans were extremophiles. Indeed, some archaea survive high temperatures, often above 100 °C (212 °F), as found in geysers, black smokers, and oil wells. Other common habitats include very cold habitats and highly saline, acidic, or alkaline water, but archaea include mesophiles that grow in mild conditions, in swamps and marshland, sewage, the oceans, the intestinal tract of animals, and soils.

Archaea that grow in the hot water of the Morning Glory Hot Spring
in YellowstoneNational Park produce a bright colour.

Extremophile archaea are members of four main physiological groups. These are the halophiles, thermophiles, alkaliphiles, and acidophiles. These groups are not comprehensive or phylum-specific, nor are they mutually exclusive, since some archaea belong to several groups. Nonetheless, they are a useful starting point for classification.

Halophiles, including the genus Halobacterium, live in extremely saline environments such as salt lakes and outnumber their bacterial counterparts at salinities greater than 20–25%. Thermophiles grow best at temperatures above 45 °C (113 °F), in places such as hot springs; hyperthermophilic archaea grow optimally at temperatures greater than 80 °C (176 °F). The archaeal Methanopyrus kandleri Strain 116 can even reproduce at 122 °C (252 °F), the highest recorded temperature of any organism.

Other archaea exist in very acidic or alkaline conditions. For example, one of the most extreme archaean acidophiles is Picrophilus torridus, which grows at pH 0, which is equivalent to thriving in 1.2 molar sulfuric acid.

This resistance to extreme environments has made archaea the focus of speculation about the possible properties of extraterrestrial life. Some extremophile habitats are not dissimilar to those on Mars, leading to the suggestion that viable microbes could be transferred between planets in meteorites.

Recently, several studies have shown that archaea exist not only in mesophilic and thermophilic environments but are also present, sometimes in high numbers, at low temperatures as well. For example, archaea are common in cold oceanic environments such as polar seas. Even more significant are the large numbers of archaea found throughout the world's oceans in non-extreme habitats among the plankton community (as part of the picoplankton). Although these archaea can be present in extremely high numbers (up to 40% of the microbial biomass), almost none of these species have been isolated and studied in pure culture. Consequently, our understanding of the role of archaea in ocean ecology is rudimentary, so their full influence on global biogeochemical cycles remains largely unexplored. Some marine Crenarchaeota are capable of nitrification, suggesting these organisms may affect the oceanic nitrogen cycle, although these oceanic Crenarchaeota may also use other sources of energy. Vast numbers of archaea are also found in the sediments that cover the sea floor, with these organisms making up the majority of living cells at depths over 1 meter below the ocean bottom. It has been demonstrated that in all oceanic surface sediments (from 1000- to 10,000-m water depth), the impact of viral infection is higher on archaea than on bacteria and virus-induced lysis of archaea accounts for up to one-third of the total microbial biomass killed, resulting in the release of ~0.3 to 0.5 gigatons of carbon per year globally.

Role in Chemical Cycling

Archaea recycle elements such as carbon, nitrogen, and sulfur through their various habitats. Although these activities are vital for normal ecosystem function, archaea can also contribute to human-made changes, and even cause pollution.

Archaea carry out many steps in the nitrogen cycle. This includes both reactions that remove nitrogen from ecosystems (such as nitrate-based respiration and denitrification) as well as processes that introduce nitrogen (such as nitrate assimilation and nitrogen fixation). Researchers recently discovered archaeal involvement in ammonia oxidation reactions. These reactions are particularly important in the oceans. The archaea also appear crucial for ammonia oxidation in soils. They produce nitrite, which other microbes then oxidize to nitrate. Plants and other organisms consume the latter.

In the sulfur cycle, archaea that grow by oxidizing sulfur compounds release this element from rocks, making it available to other organisms, but the archaea that do this, such as Sulfolobus, produce sulfuric acid as a waste product, and the growth of these organisms in abandoned mines can contribute to acid mine drainage and other environmental damage.

In the carbon cycle, methanogen archaea remove hydrogen and play an important role in the decay of organic matter by the populations of microorganisms that act as decomposers in anaerobic ecosystems, such as sediments, marshes, and sewage-treatment works.

Interactions with other Organisms

Methanogenic archaea form a symbiosis with termites.

The well-characterized interactions between archaea and other organisms are either mutual or commensal. There are no clear examples of known archaeal pathogens or parasites, but some species of methanogens have been suggested to be involved in infections in the mouth, and Nanoarchaeum equitans may be a parasite of another species of archaea, since it only survives and reproduces within the cells of the Crenarchaeon Ignicoccus hospitalis, and appears to offer no benefit to its host. Connections between archaeal cells can also be found between the Archaeal Richmond Mine Acidophilic Nanoorganisms (ARMAN) and another species of archaea called Thermoplasmatales, within acid mine drainage biofilms. Although the nature of this relationship is unknown, it is distinct from that of Nanarchaeaum–Ignicoccus in that the ultrasmall ARMAN cells are usually independent of the Thermoplasmatales cells.

Mutualism

One well-understood example of mutualism is the interaction between protozoa and methanogenic archaea in the digestive tracts of animals that digest cellulose, such as ruminants and termites. In these anaerobic environments, protozoa break down plant cellulose to obtain energy. This process releases hydrogen as a waste product, but high levels of hydrogen reduce energy production. When methanogens convert hydrogen to methane, protozoa benefit from more energy.

In anaerobic protozoa, such as Plagiopyla frontata, archaea reside inside the protozoa and consume hydrogen produced in their hydrogenosomes. Archaea also associate with larger organisms. For example, the marine archaean Cenarchaeum symbiosum lives within (is an endosymbiont of) the sponge Axinella mexicana.

Commensalism

Archaea can also be commensals, benefiting from an association without helping or harming the other organism. For example, the methanogen Methanobrevibacter smithii is by far the most common archaean in the human flora, making up about one in ten of all the prokaryotes in the human gut. In termites and in humans, these methanogens may in fact be mutualists, interacting with other microbes in the gut to aid digestion. Archaean communities also associate with a range of other organisms, such as on the surface of corals, and in the region of soil that surrounds plant roots (the rhizosphere).

Significance in Technology and Industry

Extremophile archaea, particularly those resistant either to heat or to extremes of acidity and alkalinity, are a source of enzymes that function under these harsh conditions. These enzymes have found many uses. For example, thermostable DNA polymerases, such as the Pfu DNA polymerase from Pyrococcus furiosus, revolutionized molecular biology by allowing the polymerase chain reaction to be used in research as a simple and rapid technique for cloning DNA. In industry, amylases, galactosidases and pullulanases in other species of Pyrococcus that function at over 100 °C (212 °F) allow food processing at high temperatures, such as the production of low lactose milk and whey. Enzymes from these thermophilic archaea also tend to be very stable in organic solvents, allowing their use in environmentally friendly processes in green chemistry that synthesize organic compounds. This stability makes them easier to use in structural biology. Consequently, the counterparts of bacterial or eukaryotic enzymes from extremophile archaea are often used in structural studies.

In contrast to the range of applications of archaean enzymes, the use of the organisms themselves in biotechnology is less developed. Methanogenic archaea are a vital part of sewage treatment, since they are part of the community of microorganisms that carry out anaerobic digestion and produce biogas. In mineral processing, acidophilic archaea display promise for the extraction of metals from ores, including gold, cobalt and copper.

Archaea host a new class of potentially useful antibiotics. A few of these archaeocins have been characterized, but hundreds more are believed to exist, especially within Haloarchaea and Sulfolobus. These compounds differ in structure from bacterial antibiotics, so they may have novel modes of action. In addition, they may allow the creation of new selectable markers for use in archaeal molecular biology.

Chromista

Chromista is a eukaryotic kingdom, probably polyphyletic. It includes all algae whose chloroplasts contain chlorophylls a and c, as well as various colorless forms that are closely related to them. As it is assumed the last common ancestor already possessed chloroplasts of red algal origin, the non-photosynthetic forms evolved from ancestors able to perform photosynthesis. These chloroplasts are surrounded by four membranes, and are believed to have been acquired from some red algae.

Groups

Chromista has been defined in different ways at different times. The name Chromista was first introduced by Cavalier-Smith in 1981; the earlier names Chromophyta, Chromobiota and Chromobionta correspond to roughly the same group.

It has been described as consisting of three different groups:

- Heterokonts or stramenopiles: brown algae, diatoms, water moulds, etc.

- Haptophytes.

- Cryptomonads.

In 2010, Thomas Cavalier-Smith indicated his desire to move Alveolata, Rhizaria and Heliozoa into Chromista.

Some examples of classification of the Chromista and related groups are shown below.

Chromophycées (Chadefaud, 1950)

The Chromophycées (Chadefaud, 1950), renamed Chromophycota (Chadefaud, 1960), included the current Ochrophyta (autotrophic Stramenopiles), Haptophyta (included in Chrysophyceae until Christensen, 1962), Cryptophyta, Dinophyta, Euglenophyceae and Choanoflagellida (included in Chrysophyceae until Hibberd, 1975).

Chromophyta (Christensen 1962, 1989)

The Chromophyta (Christensen 1962, 1989), defined as algae with chlorophyll c, included the current Ochrophyta (autotrophic Stramenopiles), Haptophyta, Cryptophyta, Dinophyta and Choanoflagellida. The Euglenophyceae were transferred to the Chlorophyta.

Chromophyta (Bourrelly, 1968)

The Chromophyta (Bourrelly, 1968) included the current Ochrophyta (autotrophic Stramenopiles), Haptophyta and Choanoflagellida. The Cryptophyceae and the Dinophyceae were part of Pyrrhophyta (= Dinophyta).

Chromista (Cavalier-Smith, 1981)

The Chromista (Cavalier-Smith, 1981) included the current Stramenopiles, Haptophyta and Cryptophyta.

Chromalveolata (Adl et al., 2005)

The Chromalveolata (Cavalier-Smith, 1981) included Stramenopiles, Haptophyta, Cryptophyta and Alveolata.

Chromista (Cavalier-Smith, 2010)

The Chromista (Cavalier-Smith, 2010) included SAR (Stramenopiles, Alveolata and Rhizaria) and Hacrobia (Haptista, Cryptista). A new classification of classes and phyla within Chromista was proposed by Cavalier-Smith in 2017.

Archezoa

Archezoa was a kingdom proposed by Thomas Cavalier-Smith for eukaryotes that diverged before the origin of mitochondria. At various times, the pelobionts and entamoebids (now Archamoebae), the metamonads, and the Microsporidia were included here. These groups appear near the base of eukaryotic evolution on rRNA trees. However, all these groups are now known to have developed from mitochondriate ancestors, and trees based on other genes do not support their basal placement. The kingdom Archezoa has therefore been abandoned.

Bacteria

Bacteria common noun bacteria, singular bacterium are a type of biological cell. They constitute a large domain of prokaryotic microorganisms. Typically a few micrometres in length, bacteria have a number of shapes, ranging from spheres to rods and spirals. Bacteria were among the first life forms to appear on Earth, and are present in most of its habitats. Bacteria inhabit soil, water, acidic hot springs, radioactive waste, and the deep portions of Earth's crust. Bacteria also live in symbiotic and parasitic relationships with plants and animals. Most bacteria have not been characterised, and only about 27 percent of the bacterial phyla have species that can be grown in the laboratory (specifically unculturable phyla, known as candidate phyla, make up 103 out of approximately 142 known phyla). The study of bacteria is known as bacteriology, a branch of microbiology.

Virtually all animal life on earth is dependent on bacteria for their survival as only bacteria and some archea possess the genes and enzymes necessary to synthesize vitamin B_{12}, also known as cobalamin, and provide it through the food chain. Vitamin B_{12} is a water-soluble vitamin that is involved in the metabolism of every cell of the human body. It is a cofactor in DNA synthesis, and in both fatty acid and amino acid metabolism, It is particularly important in the normal functioning of the nervous system via its role in the synthesis of myelin.

There are typically 40 million bacterial cells in a gram of soil and a million bacterial cells in a millilitre of fresh water. There are approximately 5×1030 bacteria on Earth, forming a biomass which exceeds that of all plants and animals. Bacteria are vital in many stages of the nutrient cycle by recycling nutrients such as the fixation of nitrogen from the atmosphere. The nutrient cycle includes the decomposition of dead bodies; bacteria are responsible for the putrefaction stage in this process. In the biological communities surrounding hydrothermal vents and cold seeps, extremophile bacteria provide the nutrients needed to sustain life by converting dissolved compounds, such as hydrogen sulphide and methane, to energy. Data reported by researchers in October 2012 and published in March 2013 suggested that bacteria thrive in the Mariana Trench, which, with a depth of up to 11 kilometres, is the deepest known part of the oceans. Other researchers reported related studies that microbes thrive inside rocks up to 580 metres below the sea floor under 2.6 kilometres of ocean off the coast of the northwestern United States. According to one of the researchers, "You can find microbes everywhere—they're extremely adaptable to conditions, and survive wherever they are."

The famous notion that bacterial cells in the human body outnumber human cells by a factor of 10:1 has been debunked. There are approximately 39 trillion bacterial cells in the human microbiota as personified by a "reference" 70 kg male 170 cm tall, whereas there are 30 trillion human cells in the body. This means that although they do have the upper hand in actual numbers, it is only by 30%, and not 900%.

The largest number exist in the gut flora, and a large number on the skin. The vast majority of the bacteria in the body are rendered harmless by the protective effects of the immune system, though many are beneficial, particularly in the gut flora. However several species of bacteria are pathogenic and cause infectious diseases, including cholera, syphilis, anthrax, leprosy, and bubonic plague. The most common fatal bacterial diseases are respiratory infections, with tuberculosis alone killing about 2 million people per year, mostly in sub-Saharan Africa. In developed countries, antibiotics are used to treat bacterial infections and are also used in farming, making antibiotic resistance a growing problem. In industry, bacteria are important in sewage treatment and the breakdown of oil spills, the production of cheese and yogurt through fermentation, the recovery of gold, palladium, copper and other metals in the mining sector, as well as in biotechnology, and the manufacture of antibiotics and other chemicals.

Once regarded as plants constituting the class Schizomycetes, bacteria are now classified as prokaryotes. Unlike cells of animals and other eukaryotes, bacterial cells do not contain a nucleus and rarely harbour membrane-bound organelles. Although the term bacteria traditionally included all prokaryotes, the scientific classification changed after the discovery in the 1990s that prokaryotes consist of two very different groups of organisms that evolved from an ancient common ancestor. These evolutionary domains are called Bacteria and Archaea.

Origin and Early Evolution

The ancestors of modern bacteria were unicellular microorganisms that were the first forms of life to appear on Earth, about 4 billion years ago. For about 3 billion years, most organisms were microscopic, and bacteria and archaea were the dominant forms of life. Although bacterial fossils exist, such as stromatolites, their lack of distinctive morphology prevents them from being used to examine the history of bacterial evolution, or to date the time of origin of a particular bacterial species. However, gene sequences can be used to reconstruct the bacterial phylogeny, and these studies indicate that bacteria diverged first from the archaeal/eukaryotic lineage. The most recent common ancestor of bacteria and archaea was probably a hyperthermophile that lived about 2.5 billion–3.2 billion years ago.

Bacteria were also involved in the second great evolutionary divergence, that of the archaea and eukaryotes. Here, eukaryotes resulted from the entering of ancient bacteria into endo-symbiotic associations with the ancestors of eukaryotic cells, which were themselves possibly related to the Archaea. This involved the engulfment by proto-eukaryotic cells of alphaproteobacterial symbionts to form either mitochondria or hydrogenosomes, which are still found in all known Eukarya (sometimes in highly reduced form, e.g. in ancient "amitochondrial" protozoa). Later, some eukaryotes that already contained mitochondria also engulfed cyanobacteria-like organisms, leading to the formation of chloroplasts in algae and plants. This is known as primary endosymbiosis.

In July 2018, scientists reported that the earliest life on land may have been bacteria living on land 3.22 billion years ago.

Morphology

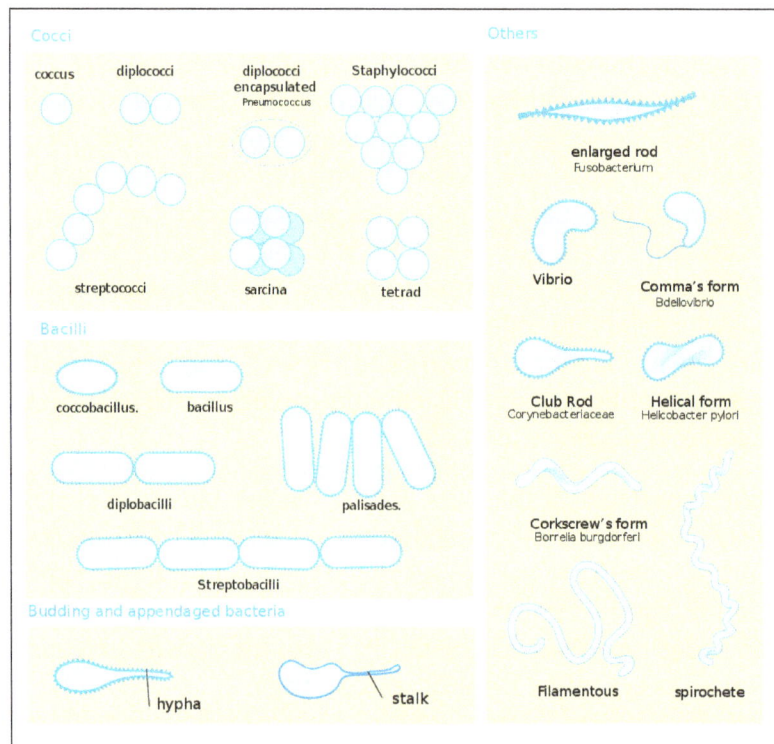

Bacteria display many cell morphologies and arrangements.

Bacteria display a wide diversity of shapes and sizes, called morphologies. Bacterial cells are about one-tenth the size of eukaryotic cells and are typically 0.5–5.0 micrometres in length.

However, a few species are visible to the unaided eye—for example, Thiomargarita namibiensis is up to half a millimetre long and Epulopiscium fishelsoni reaches 0.7 mm. Among the smallest bacteria are members of the genus Mycoplasma, which measure only 0.3 micrometres, as small as the largest viruses. Some bacteria may be even smaller, but these ultramicrobacteria are not well-studied.

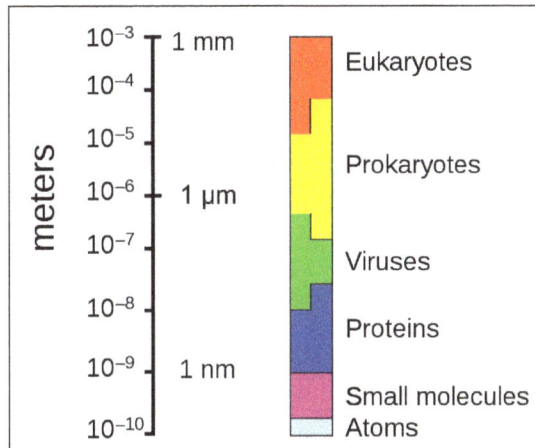

The range of sizes shown by prokaryotes, relative to those of other organisms and biomolecules.

Most bacterial species are either spherical, called cocci. Some bacteria, called vibrio, are shaped like slightly curved rods or comma-shaped; others can be spiral-shaped, called spirilla, or tightly coiled, called spirochaetes. A small number of other unusual shapes have been described, such as star-shaped bacteria. This wide variety of shapes is determined by the bacterial cell wall and cytoskeleton, and is important because it can influence the ability of bacteria to acquire nutrients, attach to surfaces, swim through liquids and escape predators.

Many bacterial species exist simply as single cells, others associate in characteristic patterns: Neisseria form diploids (pairs), Streptococcus form chains, and Staphylococcus group together in "bunch of grapes" clusters. Bacteria can also group to form larger multicellular structures, such as the elongated filaments of Actinobacteria, the aggregates of Myxobacteria, and the complex hyphae of Streptomyces. These multicellular structures are often only seen in certain conditions. For example, when starved of amino acids, Myxobacteria detect surrounding cells in a process known as quorum sensing, migrate towards each other, and aggregate to form fruiting bodies up to 500 micrometres long and containing approximately 100,000 bacterial cells. In these fruiting bodies, the bacteria perform separate tasks; for example, about one in ten cells migrate to the top of a fruiting body and differentiate into a specialised dormant state called a myxospore, which is more resistant to drying and other adverse environmental conditions.

Bacteria often attach to surfaces and form dense aggregations called biofilms, and larger formations known as microbial mats. These biofilms and mats can range from a few micrometres in thickness to up to half a metre in depth, and may contain multiple species of bacteria, protists and archaea. Bacteria living in biofilms display a complex arrangement of cells and extracellular components, forming secondary structures, such as microcolonies, through which there are networks of channels to enable better diffusion of nutrients. In natural environments, such as soil or the surfaces of plants, the majority of bacteria are bound to surfaces in biofilms. Biofilms are also important in medicine, as these structures are often present during chronic bacterial infections or in infections of implanted medical devices, and bacteria protected within biofilms are much harder to kill than individual isolated bacteria.

Cellular Structure

Intracellular Structures

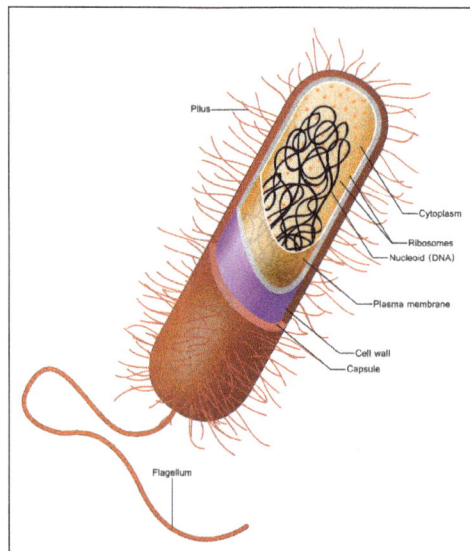

Structure and contents of a typical gram-positive bacterial cell
(seen by the fact thatonly one cell membrane is present).

The bacterial cell is surrounded by a cell membrane which is made primarily of phospholipids. This membrane encloses the contents of the cell and acts as a barrier to hold nutrients, proteins and other essential components of the cytoplasm within the cell. Unlike eukaryotic cells, bacteria usually lack large membrane-bound structures in their cytoplasm such as a nucleus, mitochondria, chloroplasts and the other organelles present in eukaryotic cells. However, some bacteria have protein-bound organelles in the cytoplasm which compartmentalize aspects of bacterial metabolism, such as the carboxysome. Additionally, bacteria have a multi-component cytoskeleton to control the localisation of proteins and nucleic acids within the cell, and to manage the process of cell division.

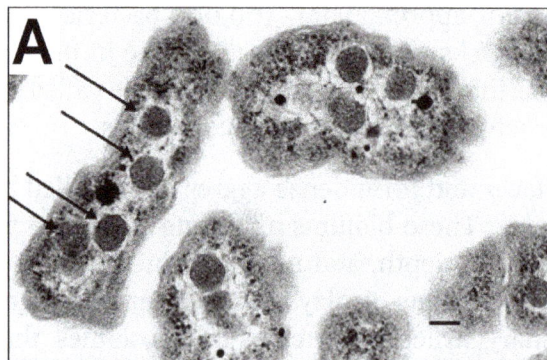

An electron micrograph of Halothiobacillus neapolitanus cells with carboxysomes inside, witharrows highlighting visible carboxysomes. Scale bars indicate 100 nm.

Many important biochemical reactions, such as energy generation, occur due to concentration gradients across membranes, creating a potential difference analogous to a battery. The general lack of internal membranes in bacteria means these reactions, such as electron transport, occur across the cell membrane between the cytoplasm and the outside of the cell or periplasm. However, in many photosynthetic bacteria the plasma membrane is highly folded and fills most of the cell with layers of light-gathering membrane. These light-gathering complexes may even form lipid-enclosed structures called chlorosomes in green sulfur bacteria.

Bacteria do not have a membrane-bound nucleus, and their genetic material is typically a single circular bacterial chromosome of DNA located in the cytoplasm in an irregularly shaped body called the nucleoid. The nucleoid contains the chromosome with its associated proteins and RNA. Like all other organisms, bacteria contain ribosomes for the production of proteins, but the structure of the bacterial ribosome is different from that of eukaryotes and Archaea.

Some bacteria produce intracellular nutrient storage granules, such as glycogen, polyphosphate, sulfur or polyhydroxyalkanoates. Certain bacterial species, such as the photosynthetic Cyanobacteria, produce internal gas vacuoles which they use to regulate their buoyancy, allowing them to move up or down into water layers with different light intensities and nutrient levels.

Extracellular Structures

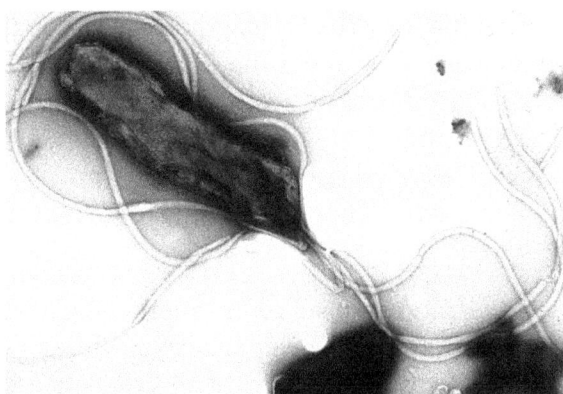

Helicobacter pylori electron micrograph, showing multiple flagella on the cell surface

Around the outside of the cell membrane is the cell wall. Bacterial cell walls are made of peptidoglycan (also called murein), which is made from polysaccharide chains cross-linked by peptides containing D-amino acids. Bacterial cell walls are different from the cell walls of plants and fungi, which are made of cellulose and chitin, respectively. The cell wall of bacteria is also distinct from that of Archaea, which do not contain peptidoglycan. The cell wall is essential to the survival of many bacteria, and the antibiotic penicillin is able to kill bacteria by inhibiting a step in the synthesis of peptidoglycan.

There are broadly speaking two different types of cell wall in bacteria, that classify bacteria into gram-positive bacteria and gram-negative bacteria. The names originate from the reaction of cells to the Gram stain, a long-standing test for the classification of bacterial species.

Gram-positive bacteria possess a thick cell wall containing many layers of peptidoglycan and teichoic acids. In contrast, gram-negative bacteria have a relatively thin cell wall consisting of a few layers of peptidoglycan surrounded by a second lipid membrane containing lipopolysaccharides and lipoproteins. Most bacteria have the gram-negative cell wall, and only the Firmicutes and Actinobacteria (previously known as the low G+C and high G+C gram-positive bacteria, respectively) have the alternative gram-positive arrangement. These differences in structure can produce differences in antibiotic susceptibility; for instance, vancomycin can kill only gram-positive bacteria and is ineffective against gram-negative pathogens, such as Haemophilus influenzae or Pseudomonas aeruginosa. Some bacteria have cell wall structures that are neither classically gram-positive or gram-negative. This includes clinically important bacteria such as Mycobacteria which have a thick peptidoglycan cell wall like a gram-positive bacterium, but also a second outer layer of lipids.

In many bacteria, an S-layer of rigidly arrayed protein molecules covers the outside of the cell. This layer provides chemical and physical protection for the cell surface and can act as a macromolecular diffusion barrier. S-layers have diverse but mostly poorly understood functions, but are known to act as virulence factors in Campylobacter and contain surface enzymes in Bacillus stearothermophilus.

Flagella are rigid protein structures, about 20 nanometres in diameter and up to 20 micrometres in length, that are used for motility. Flagella are driven by the energy released by the transfer of ions down an electrochemical gradient across the cell membrane.

Fimbriae (sometimes called "attachment pili") are fine filaments of protein, usually 2–10 nanometres in diameter and up to several micrometres in length. They are distributed over the surface of the cell, and resemble fine hairs when seen under the electron microscope. Fimbriae are believed to be involved in attachment to solid surfaces or to other cells, and are essential for the virulence of some bacterial pathogens. Pili (sing. pilus) are cellular appendages, slightly larger than fimbriae, that can transfer genetic material between bacterial cells in a process called conjugation where they are called conjugation pili or sex pili. They can also generate movement where they are called type IV pili.

Glycocalyx is produced by many bacteria to surround their cells, and varies in structural complexity: ranging from a disorganised slime layer of extracellular polymeric substances to a highly structured capsule. These structures can protect cells from engulfment by eukaryotic cells such as macrophages (part of the human immune system). They can also act as antigens and be involved in cell recognition, as well as aiding attachment to surfaces and the formation of biofilms.

The assembly of these extracellular structures is dependent on bacterial secretion systems. These transfer proteins from the cytoplasm into the periplasm or into the environment around the cell. Many types of secretion systems are known and these structures are often essential for the virulence of pathogens, so are intensively studied.

Endospores

Certain genera of gram-positive bacteria, such as Bacillus, Clostridium, Sporohalobacter, Anaerobacter, and Heliobacterium, can form highly resistant, dormant structures called endospores. Endospores develop within the cytoplasm of the cell; generally a single endospore develops in each cell. Each endospore contains a core of DNA and ribosomes surrounded by a cortex layer and protected by a multilayer rigid coat composed of peptidoglycan and a variety of proteins.

Endospores show no detectable metabolism and can survive extreme physical and chemical stresses, such as high levels of UV light, gamma radiation, detergents, disinfectants, heat, freezing, pressure, and desiccation. In this dormant state, these organisms may remain viable for millions of years, and endospores even allow bacteria to survive exposure to the vacuum and radiation in space. Endospore-forming bacteria can also cause disease: for example, anthrax can be contracted by the inhalation of Bacillus anthracis endospores, and contamination of deep puncture wounds with Clostridium tetani endospores causes tetanus.

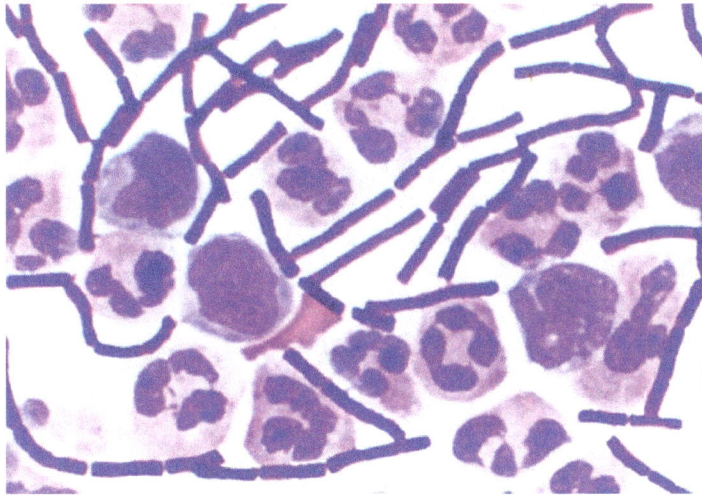

Bacillus anthracis (stained purple)
growing in cerebrospinal fluid.

Metabolism

Bacteria exhibit an extremely wide variety of metabolic types. The distribution of metabolic traits within a group of bacteria has traditionally been used to define their taxonomy, but these traits often do not correspond with modern genetic classifications. Bacterial metabolism is classified into nutritional groups on the basis of three major criteria: the source of energy, the electron donors used, and the source of carbon used for growth.

Bacteria either derive energy from light using photosynthesis (called phototrophy), or by breaking down chemical compounds using oxidation (called chemotrophy). Chemotrophs use chemical compounds as a source of energy by transferring electrons from a given electron donor to a terminal electron acceptor in a redox reaction. This reaction releases energy that can be used to drive metabolism. Chemotrophs are further divided by the types of compounds they use to transfer electrons. Bacteria that use inorganic compounds such as hydrogren, carbon monoxide, or ammonia as sources of electrons are called lithotrophs, while those that use organic compounds are called organotrophs. The compounds used to receive electrons are also used to classify bacteria: aerobic organisms use oxygen as the terminal electron acceptor, while anaerobic organisms use other compounds such as nitrate, sulfate, or carbon dioxide.

Many bacteria get their carbon from other organic carbon, called heterotrophy. Others such as cyanobacteria and some purple bacteria are autotrophic, meaning that they obtain cellular carbon by fixing carbon dioxide. In unusual circumstances, the gas methane can be used by methanotrophic bacteria as both a source of electrons and a substrate for carbon anabolism.

Nutritional Types in Bacterial Metabolism

Nutritional type	Source of energy	Source of carbon	Examples
Phototrophs	Sunlight	Organic compounds (photoheterotrophs) or carbon fixation (photoautotrophs)	Cyanobacteria, Green sulfur bacteria, Chloroflexi, or Purple bacteria
Lithotrophs	Inorganic compounds	Organic compounds (lithoheterotrophs) or carbon fixation (lithoautotrophs)	Thermodesulfobacteria, Hydrogenophilaceae, or Nitrospirae
Organotrophs	Organic compounds	Organic compounds (chemoheterotrophs) or carbon fixation (chemoautotrophs)	Bacillus, Clostridium or Enterobacteriaceae

In many ways, bacterial metabolism provides traits that are useful for ecological stability and for human society. One example is that some bacteria have the ability to fix nitrogen gas using the enzyme nitrogenase. This environmentally important trait can be found in bacteria of most metabolic types listed above. This leads to the ecologically important processes of denitrification, sulfate reduction, and acetogenesis, respectively. Bacterial metabolic processes are also important in biological responses to pollution; for example, sulfate-reducing bacteria are largely responsible for the production of the highly toxic forms of mercury (methyl- and dimethylmercury) in the environment. Non-respiratory anaerobes use fermentation to generate energy and reducing power, secreting metabolic by-products (such as ethanol in brewing) as waste. Facultative anaerobes can switch between fermentation and different terminal electron acceptors depending on the environmental conditions in which they find themselves.

Growth and Reproduction

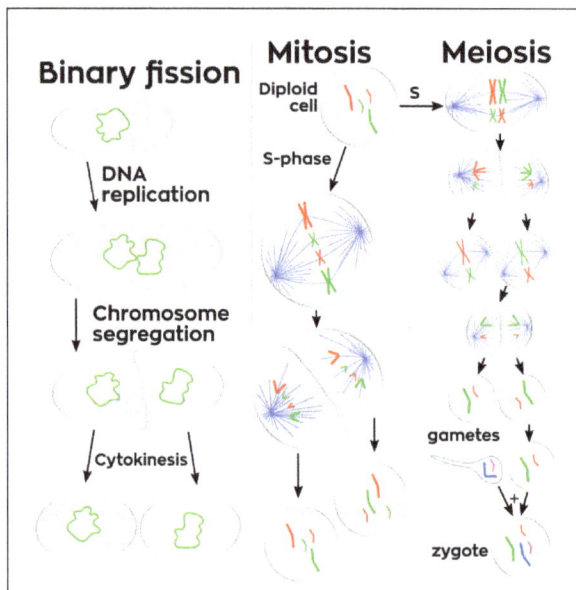

Many bacteria reproduce through binary fission, which is compared to mitosis and meiosis in this image.

Unlike in multicellular organisms, increases in cell size (cell growth) and reproduction by cell division are tightly linked in unicellular organisms. Bacteria grow to a fixed size and then reproduce through binary fission, a form of asexual reproduction. Under optimal conditions, bacteria can grow and divide extremely rapidly, and bacterial populations can double as quickly as every 9.8 minutes. In cell division, two identical clone daughter cells are produced. Some bacteria, while still reproducing asexually, form more complex reproductive structures that help disperse the newly formed daughter cells. Examples include fruiting body formation by Myxobacteria and aerial hyphae formation by Streptomyces, or budding. Budding involves a cell forming a protrusion that breaks away and produces a daughter cell.

In the laboratory, bacteria are usually grown using solid or liquid media. Solid growth media, such as agar plates, are used to isolate pure cultures of a bacterial strain. However, liquid growth media are used when measurement of growth or large volumes of cells are required. Growth in stirred liquid media occurs as an even cell suspension, making the cultures easy to divide and transfer, although isolating single bacteria from liquid media is difficult. The use of selective media (media with specific nutrients added or deficient, or with antibiotics added) can help identify specific organisms.

Most laboratory techniques for growing bacteria use high levels of nutrients to produce large amounts of cells cheaply and quickly. However, in natural environments, nutrients are limited, meaning that bacteria cannot continue to reproduce indefinitely. This nutrient limitation has led the evolution of different growth strategies. Some organisms can grow extremely rapidly when nutrients become available, such as the formation of algal (and cyanobacterial) blooms that often occur in lakes during the summer. Other organisms have adaptations to harsh environments, such as the production of multiple antibiotics by Streptomyces that inhibit the growth of competing microorganisms. In nature, many organisms live in communities (e.g., biofilms) that may allow for increased supply of nutrients and protection from environmental stresses. These relationships can be essential for growth of a particular organism or group of organisms (syntrophy).

Bacterial growth follows four phases. When a population of bacteria first enter a high-nutrient environment that allows growth, the cells need to adapt to their new environment. The first phase of growth is the lag phase, a period of slow growth when the cells are adapting to the high-nutrient environment and preparing for fast growth. The lag phase has high biosynthesis rates, as proteins necessary for rapid growth are produced. The second phase of growth is the logarithmic phase, also known as the exponential phase. The log phase is marked by rapid exponential growth. The rate at which cells grow during this phase is known as the growth rate (k), and the time it takes the cells to double is known as the generation time (g). During log phase, nutrients are metabolised at maximum speed until one of the nutrients is depleted and starts limiting growth. The third phase of growth is the stationary phase and is caused by depleted nutrients. The cells reduce their metabolic activity and consume non-essential cellular proteins. The stationary phase is a transition from rapid growth to a stress response state and there is increased expression of genes involved in DNA repair, antioxidant metabolism and nutrient transport. The final phase is the death phase where the bacteria run out of nutrients and die.

Genetics

Most bacteria have a single circular chromosome that can range in size from only 160,000 base pairs in the endosymbiotic bacteria Carsonella ruddii, to 12,200,000 base pairs (12.2 Mbp) in the soil-dwelling bacteria Sorangium cellulosum. There are many exceptions to this, for example some

Streptomyces and Borrelia species contain a single linear chromosome, while some Vibrio species contain more than one chromosome. Bacteria can also contain plasmids, small extra-chromosomal DNAs that may contain genes for various useful functions such as antibiotic resistance, metabolic capabilities, or various virulence factors.

Bacteria genomes usually encode a few hundred to a few thousand genes. The genes in bacterial genomes are usually a single continuous stretch of DNA and although several different types of introns do exist in bacteria, these are much rarer than in eukaryotes.

Bacteria, as asexual organisms, inherit an identical copy of the parent's genomes and are clonal. However, all bacteria can evolve by selection on changes to their genetic material DNA caused by genetic recombination or mutations. Mutations come from errors made during the replication of DNA or from exposure to mutagens. Mutation rates vary widely among different species of bacteria and even among different clones of a single species of bacteria. Genetic changes in bacterial genomes come from either random mutation during replication or "stress-directed mutation", where genes involved in a particular growth-limiting process have an increased mutation rate.

Some bacteria also transfer genetic material between cells. This can occur in three main ways. First, bacteria can take up exogenous DNA from their environment, in a process called transformation. Many bacteria can naturally take up DNA from the environment, while others must be chemically altered in order to induce them to take up DNA. The development of competence in nature is usually associated with stressful environmental conditions, and seems to be an adaptation for facilitating repair of DNA damage in recipient cells. The second way bacteria transfer genetic material is by transduction, when the integration of a bacteriophage introduces foreign DNA into the chromosome. Many types of bacteriophage exist, some simply infect and lyse their host bacteria, while others insert into the bacterial chromosome. Bacteria resist phage infection through restriction modification systems that degrade foreign DNA, and a system that uses CRISPR sequences to retain fragments of the genomes of phage that the bacteria have come into contact with in the past, which allows them to block virus replication through a form of RNA interference. The third method of gene transfer is conjugation, whereby DNA is transferred through direct cell contact. In ordinary circumstances, transduction, conjugation, and transformation involve transfer of DNA between individual bacteria of the same species, but occasionally transfer may occur between individuals of different bacterial species and this may have significant consequences, such as the transfer of antibiotic resistance. In such cases, gene acquisition from other bacteria or the environment is called horizontal gene transfer and may be common under natural conditions.

Behaviour

Movement

Many bacteria are motile and can move using a variety of mechanisms. The best studied of these are flagella, long filaments that are turned by a motor at the base to generate propeller-like movement. The bacterial flagellum is made of about 20 proteins, with approximately another 30 proteins required for its regulation and assembly. The flagellum is a rotating structure driven by a reversible motor at the base that uses the electrochemical gradient across the membrane for power.

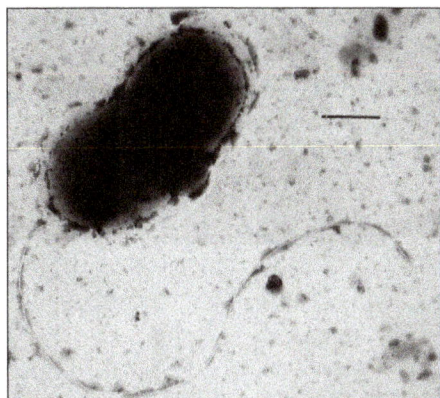

Transmission electron micrograph of Desulfovibrio vulgaris showing a single flagellum at one end of the cell. Scale bar is 0.5 micrometers long.

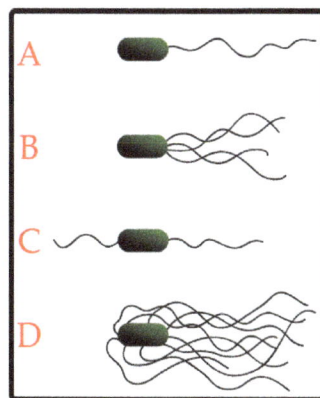

The different arrangements of bacterial flagella: A-Monotrichous; B-Lophotrichous; C-Amphitrichous; D-Peritrichous.

Bacteria can use flagella in different ways to generate different kinds of movement. Many bacteria (such as E. coli) have two distinct modes of movement: forward movement (swimming) and tumbling. The tumbling allows them to reorient and makes their movement a three-dimensional random walk. Bacterial species differ in the number and arrangement of flagella on their surface; some have a single flagellum (monotrichous), a flagellum at each end (amphitrichous), clusters of flagella at the poles of the cell (lophotrichous), while others have flagella distributed over the entire surface of the cell (peritrichous). The flagella of a unique group of bacteria, the spirochaetes, are found between two membranes in the periplasmic space. They have a distinctive helical body that twists about as it moves.

Two other types of bacterial motion are called twitching motility that relies on a structure called the type IV pilus, and gliding motility, that uses other mechanisms. In twitching motility, the rod-like pilus extends out from the cell, binds some substrate, and then retracts, pulling the cell forward.

Motile bacteria are attracted or repelled by certain stimuli in behaviours called taxes: these include chemotaxis, phototaxis, energy taxis, and magnetotaxis. In one peculiar group, the myxobacteria, individual bacteria move together to form waves of cells that then differentiate to form fruiting bodies containing spores. The myxobacteria move only when on solid surfaces, unlike E. coli, which is motile in liquid or solid media.

Several Listeria and Shigella species move inside host cells by usurping the cytoskeleton, which is normally used to move organelles inside the cell. By promoting actin polymerisation at one pole of their cells, they can form a kind of tail that pushes them through the host cell's cytoplasm.

Communication

A few bacteria have chemical systems that generate light. This bioluminescence often occurs in bacteria that live in association with fish, and the light probably serves to attract fish or other large animals.

Bacteria often function as multicellular aggregates known as biofilms, exchanging a variety of molecular signals for inter-cell communication, and engaging in coordinated multicellular behaviour.

The communal benefits of multicellular cooperation include a cellular division of labour, accessing resources that cannot effectively be used by single cells, collectively defending against antagonists, and optimising population survival by differentiating into distinct cell types. For example, bacteria in biofilms can have more than 500 times increased resistance to antibacterial agents than individual "planktonic" bacteria of the same species.

One type of inter-cellular communication by a molecular signal is called quorum sensing, which serves the purpose of determining whether there is a local population density that is sufficiently high that it is productive to invest in processes that are only successful if large numbers of similar organisms behave similarly, as in excreting digestive enzymes or emitting light.

Quorum sensing allows bacteria to coordinate gene expression, and enables them to produce, release and detect autoinducers or pheromones which accumulate with the growth in cell population.

Classification and Identification

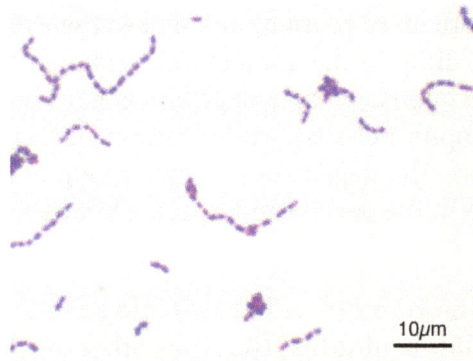

Streptococcus mutans visualised with a Gram stain.

Classification seeks to describe the diversity of bacterial species by naming and grouping organisms based on similarities. Bacteria can be classified on the basis of cell structure, cellular metabolism or on differences in cell components, such as DNA, fatty acids, pigments, antigens and quinones. While these schemes allowed the identification and classification of bacterial strains, it was unclear whether these differences represented variation between distinct species or between strains of the same species. This uncertainty was due to the lack of distinctive structures in most bacteria, as well as lateral gene transfer between unrelated species. Due to lateral gene transfer, some closely related bacteria can have very different morphologies and metabolisms. To overcome this uncertainty, modern bacterial classification emphasises molecular systematics, using genetic techniques such as guanine cytosine ratio determination, genome-genome hybridisation, as well as sequencing genes that have not undergone extensive lateral gene transfer, such as the rRNA gene. The International Committee on Systematic Bacteriology (ICSB) maintains international rules for the naming of bacteria and taxonomic categories and for the ranking of them in the International Code of Nomenclature of Bacteria.

The term "bacteria" was traditionally applied to all microscopic, single-cell prokaryotes. However, molecular systematics showed prokaryotic life to consist of two separate domains, originally called Eubacteria and Archaebacteria, but now called Bacteria and Archaea that evolved independently from an ancient common ancestor. The archaea and eukaryotes are more closely related to each other than either is to the bacteria. These two domains, along with Eukarya, are the basis of the three-domain system, which is currently the most widely used classification system in microbiology. However, due to the relatively recent introduction of molecular systematics and a rapid increase in the number of genome sequences that are available, bacterial classification remains a changing and expanding field. For example, a few biologists argue that the Archaea and Eukaryotes evolved from gram-positive bacteria.

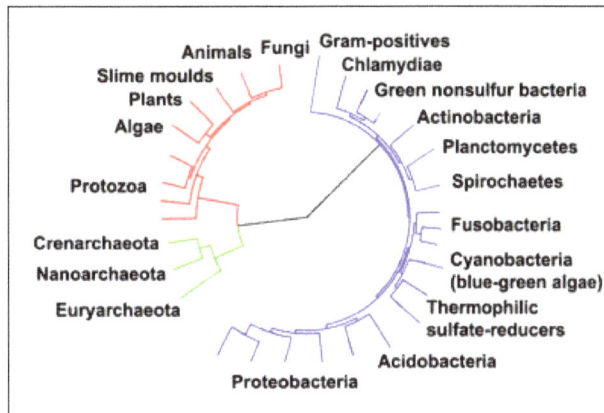

Phylogenetic tree showing the diversity of bacteria, compared to other organisms.
Eukaryotes are coloured red, archaea green and bacteria blue.

The identification of bacteria in the laboratory is particularly relevant in medicine, where the correct treatment is determined by the bacterial species causing an infection. Consequently, the need to identify human pathogens was a major impetus for the development of techniques to identify bacteria.

The Gram stain, developed in 1884 by Hans Christian Gram, characterises bacteria based on the structural characteristics of their cell walls. The thick layers of peptidoglycan in the "gram-positive" cell wall stain purple, while the thin "gram-negative" cell wall appears pink. By combining morphology and Gram-staining, most bacteria can be classified as belonging to one of four groups (gram-positive cocci, gram-positive bacilli, gram-negative cocci and gram-negative bacilli). Some organisms are best identified by stains other than the Gram stain, particularly mycobacteria or Nocardia, which show acid-fastness on Ziehl–Neelsen or similar stains. Other organisms may need to be identified by their growth in special media, or by other techniques, such as serology.

Culture techniques are designed to promote the growth and identify particular bacteria, while restricting the growth of the other bacteria in the sample. Often these techniques are designed for specific specimens; for example, a sputum sample will be treated to identify organisms that cause pneumonia, while stool specimens are cultured on selective media to identify organisms that cause diarrhoea, while preventing growth of non-pathogenic bacteria. Specimens that are normally sterile, such as blood, urine or spinal fluid, are cultured under conditions designed to grow all possible organisms. Once a pathogenic organism has been isolated, it can be further characterised by its morphology, growth patterns (such as aerobic or anaerobic growth), patterns of hemolysis, and staining.

As with bacterial classification, identification of bacteria is increasingly using molecular methods. Diagnostics using DNA-based tools, such as polymerase chain reaction, are increasingly popular due to their specificity and speed, compared to culture-based methods. These methods also allow the detection and identification of "viable but nonculturable" cells that are metabolically active but non-dividing. However, even using these improved methods, the total number of bacterial species is not known and cannot even be estimated with any certainty. Following present classification, there are a little less than 9,300 known species of prokaryotes, which includes bacteria and archaea; but attempts to estimate the true number of bacterial diversity have ranged from 107 to 109 total species—and even these diverse estimates may be off by many orders of magnitude.

Interactions with Other Organisms

Despite their apparent simplicity, bacteria can form complex associations with other organisms. These symbiotic associations can be divided into parasitism, mutualism and commensalism. Due to their small size, commensal bacteria are ubiquitous and grow on animals and plants exactly as they will grow on any other surface. However, their growth can be increased by warmth and sweat, and large populations of these organisms in humans are the cause of body odour.

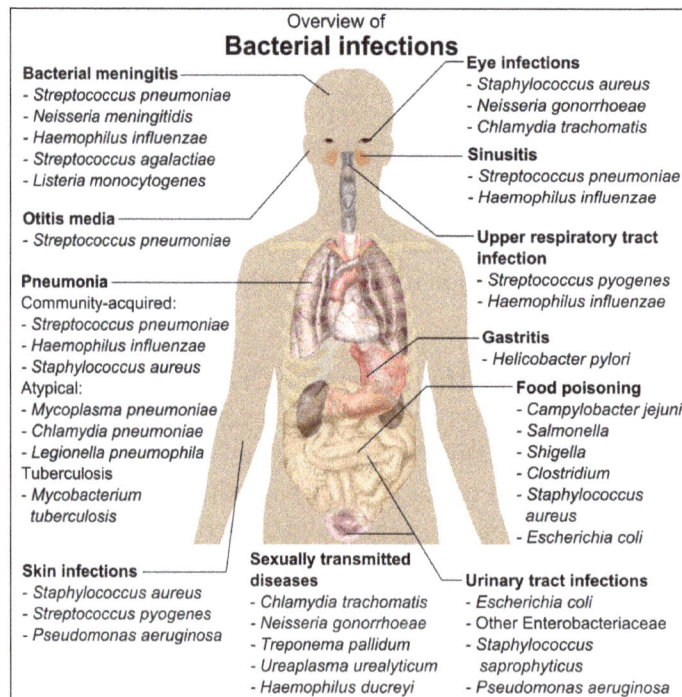

Overview of bacterial infections and main species involved.

Predators

Some species of bacteria kill and then consume other microorganisms, these species are called predatory bacteria. These include organisms such as Myxococcus xanthus, which forms swarms of cells that kill and digest any bacteria they encounter. Other bacterial predators either attach to their prey in order to digest them and absorb nutrients, such as Vampirovibrio chlorellavorus, or invade another cell and multiply inside the cytosol, such as Daptobacter. These predatory bacteria are thought to have evolved from saprophages that consumed dead microorganisms, through adaptations that allowed them to entrap and kill other organisms.

Mutualists

Certain bacteria form close spatial associations that are essential for their survival. One such mutualistic association, called interspecies hydrogen transfer, occurs between clusters of anaerobic bacteria that consume organic acids, such as butyric acid or propionic acid, and produce hydrogen, and methanogenic Archaea that consume hydrogen. The bacteria in this association are unable to consume the organic acids as this reaction produces hydrogen that accumulates in their surroundings. Only the intimate association with the hydrogen-consuming Archaea keeps the hydrogen concentration low enough to allow the bacteria to grow.

In soil, microorganisms that reside in the rhizosphere (a zone that includes the root surface and the soil that adheres to the root after gentle shaking) carry out nitrogen fixation, converting nitrogen gas to nitrogenous compounds. This serves to provide an easily absorbable form of nitrogen for many plants, which cannot fix nitrogen themselves. Many other bacteria are found as symbionts in humans and other organisms. For example, the presence of over 1,000 bacterial species in the normal human gut flora of the intestines can contribute to gut immunity, synthesise vitamins, such as folic acid, vitamin K and biotin, convert sugars to lactic acid, as well as fermenting complex undigestible carbohydrates. The presence of this gut flora also inhibits the growth of potentially pathogenic bacteria (usually through competitive exclusion) and these beneficial bacteria are consequently sold as probiotic dietary supplements.

Pathogens

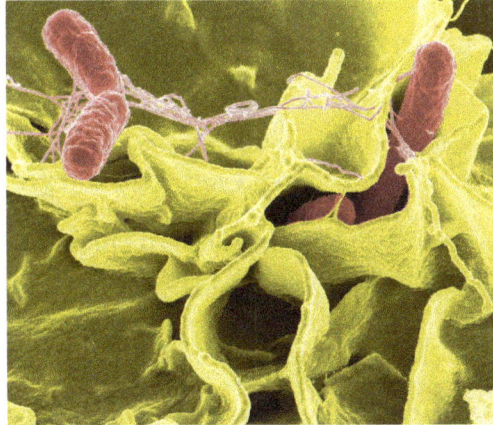

Colour-enhanced scanning electron micrograph showing Salmonella typhimurium (red) invading cultured human cells.

If bacteria form a parasitic association with other organisms, they are classed as pathogens. Pathogenic bacteria are a major cause of human death and disease and cause infections such as tetanus, typhoid fever, diphtheria, syphilis, cholera, foodborne illness, leprosy and tuberculosis. A pathogenic cause for a known medical disease may only be discovered many years after, as was the case with Helicobacter pylori and peptic ulcer disease. Bacterial diseases are also important in agriculture, with bacteria causing leaf spot, fire blight and wilts in plants, as well as Johne's disease, mastitis, salmonella and anthrax in farm animals.

Each species of pathogen has a characteristic spectrum of interactions with its human hosts. Some organisms, such as Staphylococcus or Streptococcus, can cause skin infections, pneumonia, meningitis and even overwhelming sepsis, a systemic inflammatory response producing shock, massive vasodilation and death. Yet these organisms are also part of the normal human flora and usually exist on the skin or in the nose without causing any disease at all. Other organisms invariably cause disease in humans, such as the Rickettsia, which are obligate intracellular parasites able to grow and reproduce only within the cells of other organisms. One species of Rickettsia causes typhus, while another causes Rocky Mountain spotted fever. Chlamydia, another phylum of obligate intracellular parasites, contains species that can cause pneumonia, or urinary tract infection and may be involved in coronary heart disease. Finally, some species, such as Pseudomonas aeruginosa, Burkholderia cenocepacia, and Mycobacterium avium, are opportunistic pathogens and cause disease mainly in people suffering from immunosuppression or cystic fibrosis.

Bacterial infections may be treated with antibiotics, which are classified as bacteriocidal if they kill bacteria, or bacteriostatic if they just prevent bacterial growth. There are many types of antibiotics and each class inhibits a process that is different in the pathogen from that found in the host. An example of how antibiotics produce selective toxicity are chloramphenicol and puromycin, which inhibit the bacterial ribosome, but not the structurally different eukaryotic ribosome. Antibiotics are used both in treating human disease and in intensive farming to promote animal growth, where they may be contributing to the rapid development of antibiotic resistance in bacterial populations. Infections can be prevented by antiseptic measures such as sterilising the skin prior to piercing it with the needle of a syringe, and by proper care of indwelling catheters. Surgical and dental instruments are also sterilised to prevent contamination by bacteria. Disinfectants such as bleach are used to kill bacteria or other pathogens on surfaces to prevent contamination and further reduce the risk of infection.

Significance in Technology and Industry

Bacteria, often lactic acid bacteria, such as Lactobacillus and Lactococcus, in combination with yeasts and moulds, have been used for thousands of years in the preparation of fermented foods, such as cheese, pickles, soy sauce, sauerkraut, vinegar, wine and yogurt.

The ability of bacteria to degrade a variety of organic compounds is remarkable and has been used in waste processing and bioremediation. Bacteria capable of digesting the hydrocarbons in petroleum are often used to clean up oil spills. Fertiliser was added to some of the beaches in Prince William Sound in an attempt to promote the growth of these naturally occurring bacteria after the 1989 Exxon Valdez oil spill. These efforts were effective on beaches that were not too thickly covered in oil. Bacteria are also used for the bioremediation of industrial toxic wastes. In the chemical industry, bacteria are most important in the production of enantiomerically pure chemicals for use as pharmaceuticals or agrichemicals.

Bacteria can also be used in the place of pesticides in the biological pest control. This commonly involves Bacillus thuringiensis (also called BT), a gram-positive, soil dwelling bacterium. Subspecies of this bacteria are used as a Lepidopteran-specific insecticides under trade names such as Dipel and Thuricide. Because of their specificity, these pesticides are regarded as environmentally friendly, with little or no effect on humans, wildlife, pollinators and most other beneficial insects.

Because of their ability to quickly grow and the relative ease with which they can be manipulated, bacteria are the workhorses for the fields of molecular biology, genetics and biochemistry.

By making mutations in bacterial DNA and examining the resulting phenotypes, scientists can determine the function of genes, enzymes and metabolic pathways in bacteria, then apply this knowledge to more complex organisms. This aim of understanding the biochemistry of a cell reaches its most complex expression in the synthesis of huge amounts of enzyme kinetic and gene expression data into mathematical models of entire organisms. This is achievable in some well-studied bacteria, with models of Escherichia coli metabolism now being produced and tested. This understanding of bacterial metabolism and genetics allows the use of biotechnology to bioengineer bacteria for the production of therapeutic proteins, such as insulin, growth factors, or antibodies.

Because of their importance for research in general, samples of bacterial strains are isolated and preserved in Biological Resource Centers. This ensures the availability of the strain to scientists worldwide.

PHYLUM

In biology, a phylum is a level of classification or taxonomic rank below kingdom and above class. Traditionally, in botany the term division has been used instead of phylum, although the International Code of Nomenclature for algae, fungi, and plants accepts the terms as equivalent. Depending on definitions, the animal kingdom Animalia or Metazoa contains approximately 35 phyla, the plant kingdom Plantae contains about 14, and the fungus kingdom Fungi contains about 8 phyla. Current research in phylogenetics is uncovering the relationships between phyla, which are contained in larger clades, like Ecdysozoa and Embryophyta.

Haeckel noted that species constantly evolved into new species that seemed to retain few consistent features among themselves and therefore few features that distinguished them as a group ("a self-contained unity"). "Wohl aber ist eine solche reale und vollkommen abgeschlossene Einheit die Summe aller Species, welche aus einer und derselben gemeinschaftlichen Stammform allmählig sich entwickelt haben, wie z. B. alle Wirbelthiere. Diese Summe nennen wir Stamm (Phylon)." which translates as: However, perhaps such a real and completely self-contained unity is the aggregate of all species which have gradually evolved from one and the same common original form, as, for example, all vertebrates. We name this aggregate Stamm [i.e., race] (Phylon).) In plant taxonomy, August W. Eichler classified plants into five groups named divisions, a term that remains in use today for groups of plants, algae and fungi. The definitions of zoological phyla have changed from their origins in the six Linnaean classes and the four embranchements of Georges Cuvier.

Informally, phyla can be thought of as groupings of organisms based on general specialization of body plan. At its most basic, a phylum can be defined in two ways: as a group of organisms with a certain degree of morphological or developmental similarity (the phenetic definition), or a group of organisms with a certain degree of evolutionary relatedness (the phylogenetic definition). Attempting to define a level of the Linnean hierarchy without referring to (evolutionary) relatedness is unsatisfactory, but a phenetic definition is useful when addressing questions of a morphological nature—such as how successful different body plans were.

Based on Genetic Relation

The most important objective measure in the above definitions is the "certain degree" that defines how different organisms need to be members of different phyla. The minimal requirement is that all organisms in a phylum should be clearly more closely related to one another than to any other group. Even this is problematic because the requirement depends on knowledge of organisms' relationships: as more data become available, particularly from molecular studies, we are better able to determine the relationships between groups. So phyla can be merged or split if it becomes apparent that they are related to one another or not. For example, the bearded worms were described as a new phylum (the Pogonophora) in the middle of the 20th century, but molecular work almost half a century later found them to be a group of annelids, so the phyla were merged (the bearded worms are now an annelid family). On the other hand, the highly parasitic phylum Mesozoa was divided into two phyla (Orthonectida and Rhombozoa) when it was discovered the Orthonectida are probably deuterostomes and the Rhombozoa protostomes.

This changeability of phyla has led some biologists to call for the concept of a phylum to be abandoned in favour of cladistics, a method in which groups are placed on a "family tree" without any formal ranking of group size.

Based on Body Plan

A definition of a phylum based on body plan has been proposed by paleontologists Graham Budd and Sören Jensen (as Haeckel had done a century earlier). The definition was posited because extinct organisms are hardest to classify: they can be offshoots that diverged from a phylum's line before the characters that define the modern phylum were all acquired. By Budd and Jensen's definition, a phylum is defined by a set of characters shared by all its living representatives.

This approach brings some small problems—for instance, ancestral characters common to most members of a phylum may have been lost by some members. Also, this definition is based on an arbitrary point of time: the present. However, as it is character based, it is easy to apply to the fossil record. A greater problem is that it relies on a subjective decision about which groups of organisms should be considered as phyla.

The approach is useful because it makes it easy to classify extinct organisms as "stem groups" to the phyla with which they bear the most resemblance, based only on the taxonomically important similarities. However, proving that a fossil belongs to the crown group of a phylum is difficult, as it must display a character unique to a sub-set of the crown group. Furthermore, organisms in the stem group of a phylum can possess the "body plan" of the phylum without all the characteristics necessary to fall within it. This weakens the idea that each of the phyla represents a distinct body plan.

A classification using this definition may be strongly affected by the chance survival of rare groups, which can make a phylum much more diverse than it would be otherwise.

Plants

The kingdom Plantae is defined in various ways by different biologists. All definitions include the living embryophytes (land plants), to which may be added the two green algae divisions, Chlorophyta and Charophyta, to form the clade Viridiplantae. The table below follows the influential (though contentious) Cavalier-Smith system in equating "Plantae" with Archaeplastida, a group containing Viridiplantae and the algal Rhodophyta and Glaucophyta divisions.

The definition and classification of plants at the division level also varies from source to source, and has changed progressively in recent years. Thus some sources place horsetails in division Arthrophyta and ferns in division Pteridophyta, while others place them both in Pteridophyta, as shown below. The division Pinophyta may be used for all gymnosperms (i.e. including cycads, ginkgos and gnetophytes), or for conifers alone as below.

Since the first publication of the APG system in 1998, which proposed a classification of angiosperms up to the level of orders, many sources have preferred to treat ranks higher than orders as informal clades. Where formal ranks have been provided, the traditional divisions listed below have been reduced to a very much lower level, e.g. subclasses.

Fungi

Division	Meaning	Common name	Distinguishing characteristics
Ascomycota	Bladder fungus	Ascomycetes, sac fungi	Tend to have fruiting bodies (ascocarp). Filamentous, producing hyphae separated by septa. Asexual reproduction.
Basidiomycota	Small base fungus	Basidiomycetes	Bracket fungi, toadstools, smuts and rust. Sexual reproduction.
Blastocladiomycota	Offshoot branch fungus	Blastoclads	
Chytridiomycota	Little cooking pot fungus	Chytrids	Predominantly Aquatic saprotrophic or parasitic. Have a posterior flagellum. Tend to be single celled but can also be multicellular.
Glomeromycota	Ball of yarn fungus	Glomeromycetes, AM fungi	Mainly arbuscular mycorrhizae present, terrestrial with a small presence on wetlands. Reproduction is asexual but requires plant roots.
Microsporidia	Small seeds	Microsporan	
Neocallimastigomycota	New beautiful whip fungus	Neocallimastigomycetes	Predominantly located in digestive tract of herbivorus animals. Anaerobic, terrestrial and aquatic.
Zygomycota	Pair fungus	Zygomycetes	Most are saprobes and reproduce sexually and asexually.
Total: 8			

Phylum Microsporidia is generally included in kingdom Fungi, though its exact relations remain uncertain, and it is considered a protozoan by the International Society of Protistologists. Molecular analysis of Zygomycota has found it to be polyphyletic (its members do not share an immediate ancestor), which is considered undesirable by many biologists. Accordingly, there is a proposal to abolish the Zygomycota phylum. Its members would be divided between phylum Glomeromycota and four new subphyla incertae sedis (of uncertain placement): Entomophthoromycotina, Kickxellomycotina, Mucoromycotina, and Zoopagomycotina.

Bacteria

Currently there are 29 phyla accepted by List of Prokaryotic names with Standing in Nomenclature (LPSN)

- Acidobacteria, phenotipically diverse and mostly uncultured

- Actinobacteria, High-G⁺C Gram positive species

- Aquificae, only 14 thermophilic genera, deep branching

- Armatimonadetes

- Bacteroidetes

- Caldiserica, formerly candidate division OP5, Caldisericum exile is the sole representative

- Chlamydiae, only 6 genera

- Chlorobi, only 7 genera, green sulphur bacteria

- Chloroflexi, green non-sulphur bacteria

- Chrysiogenetes, only 3 genera (Chrysiogenes arsenatis, Desulfurispira natronophila, Desulfurispirillum alkaliphilum)

- Cyanobacteria, also known as the blue-green algae

- Deferribacteres

- Deinococcus-Thermus, Deinococcus radiodurans and Thermus aquaticus are "commonly known" species of this phyla

- Dictyoglomi

- Elusimicrobia, formerly candidate division Thermite Group 1

- Fibrobacteres

- Firmicutes, Low-G+C Gram positive species, such as the spore-formers Bacilli (aerobic) and Clostridia (anaerobic)

- Fusobacteria

- Gemmatimonadetes

- Lentisphaerae, formerly clade VadinBE97

- Nitrospira

- Planctomycetes

- Proteobacteria, the most known phyla, containing species such as Escherichia coli or Pseudomonas aeruginosa

- Spirochaetes, species include Borrelia burgdorferi, which causes Lyme disease

- Synergistetes

- Tenericutes, alternatively class Mollicutes in phylum Firmicutes (notable genus: Mycoplasma)

- Thermodesulfobacteria

- Thermotogae, deep branching

- Verrucomicrobia

Archaea

Currently there are 5 phyla accepted by List of Prokaryotic names with Standing in Nomenclature (LPSN).

- Crenarchaeota, second most common archaeal phylum

- Euryarchaeota, most common archaeal phylum

- Korarchaeota

- Nanoarchaeota, ultra-small symbiotes, single known species

CLASS

In biological classification, class is a taxonomic rank, as well as a taxonomic unit, a taxon, in that rank. Other well-known ranks in descending order of size are life, domain, kingdom, phylum, order, family, genus, and species, with class fitting between phylum and order.

The class as a distinct rank of biological classification having its own distinctive name (and not just called a top-level genus (genus summum)) was first introduced by the French botanist Joseph Pitton de Tournefort in his classification of plants that appeared in his Eléments de botanique, 1694.

Insofar as a general definition of a class is available, it has historically been conceived as embracing taxa that combine a distinct grade of organization -- i.e. a 'level of complexity', measured in terms of how differentiated their organ systems are into distinct regions or sub-organs -- with a distinct type of construction, which is to say a particular layout of organ systems. This said, the composition of each class is ultimately determined by the subjective judgement of taxonomists. Often there is no exact agreement, with different taxonomists taking different positions. There are no objective rules for describing a class, but for well-known animals there is likely to be consensus.

In the first edition of his Systema Naturae (1735). Carl Linnaeus divided all three of his kingdoms of Nature (minerals, plants, and animals) into classes. Only in the animal kingdom are Linnaeus's classes similar to the classes used today; his classes and orders of plants were never intended to represent natural groups, but rather to provide a convenient "artificial key" according to his

Systema Sexuale, largely based on the arrangement of flowers. In botany, classes are now rarely discussed. Since the first publication of the APG system in 1998, which proposed a taxonomy of the flowering plants up to the level of orders, many sources have preferred to treat ranks higher than orders as informal clades. Where formal ranks have been assigned, the ranks have been reduced to a very much lower level, e.g. class Equisitopsida for the land plants, with the major divisions within the class assigned to subclasses and superorders.

The class was considered the highest level of the taxonomic hierarchy until George Cuvier's embranchements, first called Phyla by Ernst Haeckel, were introduced in the early nineteenth century.

Hierarchy of Ranks below and above the Level of Class

As for the other principal ranks, Classes can be grouped and subdivided. Here are some examples.

Name	Meaning of prefix	Example 1	Example 2	Example 3	Example 4
Superclass	super: above	Tetrapoda		Tetrapoda	
Class		Mammalia	Maxillopoda	Aves	Diplopoda
Subclass	sub: under	Theria	Thecostraca		Chilognatha
Infraclass	infra: below		Cirripedia	Neognathae	Helminthomorpha
Subterclass	subter: below, underneath				Colobognatha
Parvclass	parvus: small, unimportant			Neornithes	-

ORDER

In biological classification, the order is:

1. A taxonomic rank used in the classification of organisms and recognized by the nomenclature codes. Other well-known ranks are life, domain, kingdom, phylum, class, family, genus, and species, with order fitting in between class and family. An immediately higher rank, superorder, may be added directly above order, while suborder would be a lower rank.

2. A taxonomic unit, a taxon, in that rank. In that case the plural is orders.

Example: All owls belong to the order Strigiformes.

What does and does not belong to each order is determined by a taxonomist, as is whether a particular order should be recognized at all. Often there is no exact agreement, with different taxonomists each taking a different position. There are no hard rules that a taxonomist needs to follow in describing or recognizing an order. Some taxa are accepted almost universally, while others are recognised only rarely.

For some groups of organisms, consistent suffixes are used to denote that the rank is an order. The Latin suffix -(i)formes meaning "having the form of" is used for the scientific name of orders of birds and fishes, but not for those of mammals and invertebrates. The suffix -ales is for the name of orders of plants, fungi, and algae.

Hierarchy of Ranks

Zoology

For some clades covered by the International Code of Zoological Nomenclature, a number of additional classifications are sometimes used, although not all of these are officially recognised.

Name	Meaning of prefix	Example 1	Example 2
Magnorder	magnus: large, great, important	Boreoeutheria	
Superorder	super: above	Euarchontoglires	Parareptilia
Grandorder	grand: large	Euarchonta	
Mirorder	mirus: wonderful, strange	Primatomorpha	
Order		Primates	Procolophonomorpha
Suborder	sub: under	Haplorrhini	Procolophonia
Infraorder	infra: below	Simiiformes	Hallucicrania
Parvorder	parvus: small, unimportant	Catarrhini	

In their 1997 classification of mammals, McKenna and Bell used two extra levels between superorder and order: "grandorder" and "mirorder". Michael Novacek inserted them at the same position. Michael Benton inserted them between superorder and magnorder instead. This position was adopted by Systema Naturae 2000 and others.

Botany

In botany, the ranks of subclass and suborder are secondary ranks pre-defined as respectively above and below the rank of order. Any number of further ranks can be used as long as they are clearly defined.

The superorder rank is commonly used, with the ending -anae that was initiated by Armen Takhtajan's publications from 1966 onwards.

FAMILY

Family is one of the eight major hierarchical taxonomic ranks in Linnaean taxonomy; it is classified between order and genus. A family may be divided into subfamilies, which are intermediate ranks between the ranks of family and genus. The official family names are Latin in origin; however, popular names are often used: for example, walnut trees and hickory trees belong to the family Juglandaceae, but that family is commonly referred to as being the "walnut family".

What does or does not belong to a family—or whether a described family should be recognized at all—are proposed and determined by practicing taxonomists. There are no hard rules for describing or recognizing a family. Taxonomists often take different positions about descriptions, and there may be no broad consensus across the scientific community for some time. The publishing of new data and opinion often enables adjustments and consensus.

Nomenclature

The naming of families is codified by various international bodies using the following suffixes:

- In fungal, algal, and botanical nomenclature, the family names of plants, fungi, and algae end with the suffix "-aceae", with the exception of a small number of historic but widely used names including Compositae and Gramineae.

- In zoological nomenclature, the family names of animals end with the suffix "-idae".

Uses

Families can be used for evolutionary, palaeontological and genetic studies because they are more stable than lower taxonomic levels such as genera and species.

Mnemonic

An easily remembered mnemonic for the taxonomic hierarchy within which the term family lies is "Doctor King Phillip Came Over From German Shores."

GENUS

A genus is a taxonomic rank used in the biological classification of living and fossil organisms, as well as viruses, in biology. In the hierarchy of biological classification, genus comes above species and below family. In binomial nomenclature, the genus name forms the first part of the binomial species name for each species within the genus.

E.g. Panthera leo (lion) and Panthera onca (jaguar) are two species within the genus Panthera. Panthera is a genus within the family Felidae.

The composition of a genus is determined by a taxonomist. The standards for genus classification are not strictly codified, so different authorities often produce different classifications for genera. There are some general practices used, however, including the idea that a newly defined genus should fulfill these three criteria to be descriptively useful:

1. Monophyly – all descendants of an ancestral taxon are grouped together (i.e. phylogenetic analysis should clearly demonstrate both monophyly and validity as a separate lineage).

2. Reasonable compactness – a genus should not be expanded needlessly; and

3. Distinctness – with respect to evolutionarily relevant criteria, i.e. ecology, morphology, or biogeography; DNA sequences are a consequence rather than a condition of diverging evolutionary lineages except in cases where they directly inhibit gene flow (e.g. postzygotic barriers).

Moreover, genera should be composed of phylogenetic units of the same kind as other (analogous) genera.

Use

The scientific name (or the scientific epithet) of a genus is also called the generic name; it is always capitalised. It plays a fundamental role in binomial nomenclature, the system of naming organisms, where it is combined with the scientific name of a species.

Use in Nomenclature

The rules for the scientific names of organisms are laid down in the Nomenclature Codes, which allow each species a single unique name that, for "animals" (including protists), "plants" (also including algae and fungi) and prokaryotes (Bacteria and Archaea), is Latin and binomial in form; this contrasts with common or vernacular names, which are non-standardized, can be non-unique, and typically also vary by country and language of usage.

Except for viruses, the standard format for a species name comprises the generic name, indicating the genus to which the species belongs, followed by the specific epithet, which (within that genus) is unique to the species. For example, the gray wolf's scientific name is Canis lupus, with Canis (Lat. "dog") being the generic name shared by the wolf's close relatives and lupus (Lat. "wolf") being the specific name particular to the wolf. A botanical example would be Hibiscus arnottianus, a particular species of the genus Hibiscus native to Hawaii. The specific name is written in lower-case and may be followed by subspecies names in zoology or a variety of infraspecific names in botany.

When the generic name is already known from context, it may be shortened to its initial letter, for example C. lupus in place of Canis lupus. Where species are further subdivided, the generic name (or its abbreviated form) still forms the leading portion of the scientific name, for example Canis lupus familiaris for the domestic dog (when considered a subspecies of the gray wolf) in zoology, or as a botanical example, Hibiscus arnottianus ssp. immaculatus. Also, as visible in the above examples, the Latinised portions of the scientific names of genera and their included species (and infraspecies, where applicable) are, by convention, written in italics.

The scientific names of virus species are descriptive, not binomial in form, and may or may not incorporate an indication of their containing genus; for example the virus species "Salmonid herpesvirus 1", "Salmonid herpesvirus 2" and "Salmonid herpesvirus 3" are all within the genus Salmonivirus, however the genus to which the species with the formal names "Everglades virus" and "Ross River virus" are assigned is Alphavirus.

As with scientific names at other ranks, in all groups other than viruses, names of genera may be cited with their authorities, typically in the form "author, year" in zoology, and "standard abbreviated author name" in botany. Thus in the examples above, the genus Canis would be cited in full as "Canis Linnaeus, 1758" (zoological usage), while Hibiscus, also first established by Linnaeus but in 1753, is simply "Hibiscus L." (botanical usage).

The Type Concept

Each genus should have a designated type, although in practice there is a backlog of older names without one. In zoology, this is the type species and the generic name is permanently associated with the type specimen of its type species. Should the specimen turn out to be assignable to another genus, the generic name linked to it becomes a junior synonym and the remaining taxa in the former genus need to be reassessed.

Categories of Generic Name

In zoological usage, taxonomic names, including those of genera, are classified as "available" or "unavailable". Available names are those published in accordance with the International Code of Zoological Nomenclature and not otherwise suppressed by subsequent decisions of the International Commission on Zoological Nomenclature (ICZN); the earliest such name for any taxon (for example, a genus) should then be selected as the "valid" (i.e., current or accepted) name for the taxon in question.

Consequently, there will be more available names than valid names at any point in time, which names are currently in use depending on the judgement of taxonomists in either combining taxa described under multiple names, or splitting taxa which may bring available names previously treated as synonyms back into use. "Unavailable" names in zoology comprise names that either were not published according to the provisions of the ICZN Code, or have subsequently been suppressed, e.g., incorrect original or subsequent spellings, names published only in a thesis, and generic names published after 1930 with no type species indicated.

In botany, similar concepts exist but with different labels. The botanical equivalent of zoology's "available name" is a validly published name. An invalidly published name is a nomen invalidum or nom. inval.; a rejected name is a nomen rejiciendum or nom. rej.; a later homonym of a validly published name is a nomen illegitimum or nom. illeg.; for a full list refer the International Code of Nomenclature for algae, fungi, and plants (ICNafp) and the work cited above by Hawksworth, 2010. In place of the "valid taxon" in zoology, the nearest equivalent in botany is "correct name" or "current name" which can, again, differ or change with alternative taxonomic treatments or new information that results in previously accepted genera being combined or split.

Prokaryote and virus Codes of Nomenclature also exist which serve as a reference for designating currently accepted genus names as opposed to others which may be either reduced to synonymy, or, in the case of prokaryotes, relegated to a status of "names without standing in prokaryotic nomenclature".

An available (zoological) or validly published (botanical) name that has been historically applied to a genus but is not regarded as the accepted (current/valid) name for the taxon is termed a synonym; some authors also include unavailable names in lists of synonyms as well as available names, such as misspellings, names previously published without fulfilling all of the requirements of the relevant nomenclatural Code, and rejected or suppressed names.

A particular genus name may have zero to many synonyms, the latter case generally if the genus has been known for a long time and redescribed as new by a range of subsequent workers, or if a range of genera previously considered separate taxa have subsequently been consolidated into one. For example, the World Register of Marine Species presently lists 8 genus-level synonyms for the sperm whale genus Physeter Linnaeus, 1758, and 13 for the bivalve genus Pecten O.F. Müller, 1776.

Identical Names (Homonyms)

Within the same kingdom, one generic name can apply to one genus only. However, many names have been assigned (usually unintentionally) to two or more different genera. For example, the platypus belongs to the genus Ornithorhynchus although George Shaw named it Platypus in 1799 (these two names are thus synonyms). However, the name Platypus had already been given to a group of ambrosia beetles by Johann Friedrich Wilhelm Herbst in 1793. A name that means two different things is a homonym. Since beetles and platypuses are both members of the kingdom Animalia, the name could not be used for both. Johann Friedrich Blumenbach published the replacement name Ornithorhynchus in 1800.

However, a genus in one kingdom is allowed to bear a scientific name that is in use as a generic name (or the name of a taxon in another rank) in a kingdom that is governed by a different nomenclature code. Names with the same form but applying to different taxa are called "homonyms". Although this is discouraged by both the International Code of Zoological Nomenclature and the International Code of Nomenclature for algae, fungi, and plants, there are some five thousand such names in use in more than one kingdom. For instance,

- Anura is the name of the order of frogs but also is the name of a non-current genus of plants;
- Aotus is the generic name of both golden peas and night monkeys;
- Oenanthe is the generic name of both wheatears and water dropworts;
- Prunella is the generic name of both accentors and self-heal;
- Proboscidea is the order of elephants and the genus of devil's claws;
- The name of the genus Paramecia (an extinct red alga) is also the plural of the name of the genus Paramecium (which is in the SAR supergroup), which can also lead to confusion.

A list of generic homonyms (with their authorities), including both available (validly published) and selected unavailable names, has been compiled by the Interim Register of Marine and Non-marine Genera (IRMNG).

Use in Higher Classifications

The type genus forms the base for higher taxonomic ranks, such as the family name Canidae ("Canids") based on Canis. However, this does not typically ascend more than one or two levels: the order to which dogs and wolves belong is Carnivora ("Carnivores").

Numbers of Accepted Genera

The numbers of either accepted, or all published genus names is not known precisely although the latter value has been estimated by Rees et al., 2017 at approximately 510,000 as at end 2016, increasing at some 2,500 per year. "Official" registers of taxon names at all ranks, including genera, exist for a few groups only such as viruses and prokaryotes, while for others there are compendia with no "official" standing such as Index Fungorum for Fungi, Index Nominum Algarum and AlgaeBase for algae, Index Nominum Genericorum and the International Plant Names Index for plants in general, and ferns through angiosperms, respectively, and Nomenclator Zoologicus and the Index to Organism Names for zoological names.

A deduplicated list of genus names covering all taxonomic groups, compiled from resources such as the above as well as other literature sources, created as the "Interim Register of Marine and

Nonmarine Genera" (IRMNG), is estimated to contain around 95% of all published names at generic level, and lists approximately 490,100 genus names in its March 2019 release. Of these, approx. 265,500 are presently flagged "accepted" (including both extant and fossil taxa), 127,500 as unaccepted for a range of reasons, and an additional 126,000 not yet assessed for taxonomic status. In this compilation, therefore, an estimate of the total number of accepted names thus lies somewhere between the values of 265,500 (if all unassessed names are in fact synonyms or otherwise unaccepted names) and 391,500 (if all unassessed names are accepted taxa) with a mean value in the order of 330,000 names. Included in the 265,500 presently "accepted", extant plus fossil genus names in the March 2019 edition of IRMNG are 188,158 genera of animals (kingdom Animalia), 21,935 Plantae (land plants and non-Chromistan algae), 10,231 Fungi, 9,989 Chromista, 1,963 Protozoa, 3,387 Prokaryotes (3,247 Bacteria plus 140 Archaea) and 851 Viruses, although totals for some groups (mainly animals, Chromista and Protozoa, where the bulk of the present unassessed names reside) will be underestimates since presently unassessed names will divide between "accepted" and "unaccepted" categories once they are further investigated.

By comparison, the 2018 annual edition of the Catalogue of Life (estimated >90% complete, for extant species in the main) contains currently 175,363 "accepted" genus names for 1,744,204 living and 59,284 extinct species, also including genus names only (no species) for some groups.

Genus Size

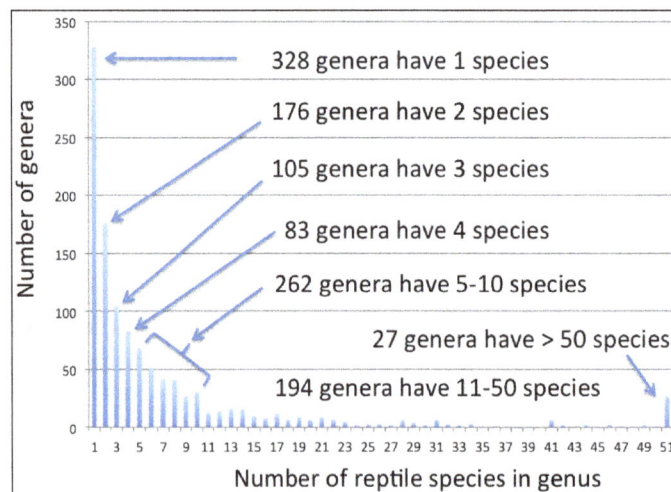

Number of reptile genera with a given number of species. Most genera have only one or a few species but a few may have hundreds. Based on data from the Reptile Database .

The number of species in genera varies considerably among taxonomic groups. For instance, among (non-avian) reptiles, which have about 1180 genera, the most (>300) have only 1 species, ~360 have between 2 and 4 species, 260 have 5-10 species, ~200 have 11-50 species, and only 27 genera have more than 50 species. However, some insect genera such as the bee genera Lasioglossum and Andrena have over 1000 species each. The largest flowering plant genus, Astragalus, contains over 3,000 species.

Which species are assigned to a genus is somewhat arbitrary. Although all species within a genus are supposed to be "similar" there are no objective criteria for grouping species into genera. There is much debate among zoologists whether large, species-rich genera should be maintained, as it is extremely difficult to come up with identification keys or even character sets that distinguish all species. Hence, many taxonomists argue in favor of breaking down large genera. For instance, the lizard genus Anolis has been suggested to be broken down into 8 or so different genera which would bring its ~400 species to smaller, more manageable subsets.

SPECIES

In biology, a species is the basic unit of classification and a taxonomic rank of an organism, as well as a unit of biodiversity. A species is often defined as the largest group of organisms in which any two individuals of the appropriate sexes or mating types can produce fertile offspring, typically by sexual reproduction. Other ways of defining species include their karyotype, DNA sequence, morphology, behaviour or ecological niche. In addition, paleontologists use the concept of the chronospecies since fossil reproduction cannot be examined. While these definitions may seem adequate, when looked at more closely they represent problematic species concepts. For example, the boundaries between closely related species become unclear with hybridisation, in a species complex of hundreds of similar microspecies, and in a ring species. Also, among organisms that reproduce only asexually, the concept of a reproductive species breaks down, and each clone is potentially a microspecies.

All species (except viruses) are given a two-part name, a "binomial". The first part of a binomial is the genus to which the species belongs. The second part is called the specific name or the specific epithet (in botanical nomenclature, also sometimes in zoological nomenclature). For example, Boa constrictor is one of four species of the genus Boa.

None of these are entirely satisfactory definitions, but scientists and conservationists need a species definition which allows them to work, regardless of the theoretical difficulties. If species were fixed and clearly distinct from one another, there would be no problem, but evolutionary processes cause species to change continually, and to grade into one another.

Species were seen from the time of Aristotle until the 18th century as fixed kinds that could be arranged in a hierarchy, the great chain of being. In the 19th century, biologists grasped that species could evolve given sufficient time. Charles Darwin's 1859 book The Origin of Species explained how species could arise by natural selection. That understanding was greatly extended in the 20th century through genetics and population ecology. Genetic variability arises from mutations and recombination, while organisms themselves are mobile, leading to geographical isolation and genetic drift with varying selection pressures. Genes can sometimes be exchanged between species by horizontal gene transfer; new species can arise rapidly through hybridisation and polyploidy; and species may become extinct for a variety of reasons. Viruses are a special case, driven by a balance of mutation and selection, and can be treated as quasispecies.

Biologists and taxonomists have made many attempts to define species, beginning from morphology and moving towards genetics. Early taxonomists such as Linnaeus had no option but to describe what they saw: this was later formalised as the typological or morphological species concept. Ernst Mayr emphasised reproductive isolation, but this, like other species concepts, is hard or even impossible to test. Later biologists have tried to refine Mayr's definition with the recognition and cohesion concepts, among others. Many of the concepts are quite similar or overlap, so they are not easy to count: the biologist R. L. Mayden recorded about 24 concepts, and the philosopher of science John Wilkins counted 26. Wilkins further grouped the species concepts into seven basic kinds of concepts: (1) agamospecies for asexual organisms (2) biospecies for reproductively isolated sexual organisms (3) ecospecies based on ecological niches (4) evolutionary species based on lineage (5) genetic species based on gene pool (6) morphospecies based on form or phenotype and (7) taxonomic species, a species as determined by a taxonomist.

Typological or Morphological Species

A typological species is a group of organisms in which individuals conform to certain fixed properties (a type), so that even pre-literate people often recognise the same taxon as do modern taxonomists. The clusters of variations or phenotypes within specimens (such as longer or shorter tails) would differentiate the species. This method was used as a "classical" method of determining species, such as with Linnaeus early in evolutionary theory. However, different phenotypes are not necessarily different species (e.g. a four-winged Drosophila born to a two-winged mother is not a different species). Species named in this manner are called morphospecies.

All adult Eurasian blue tits share the same coloration,
unmistakably identifying the morphospecies.

In the 1970s, Robert R. Sokal, Theodore J. Crovello and Peter Sneath proposed a variation on this, a phenetic species, defined as a set of organisms with a similar phenotype to each other, but a different phenotype from other sets of organisms. It differs from the morphological species concept in including a numerical measure of distance or similarity to cluster entities based on multivariate comparisons of a reasonably large number of phenotypic traits.

Recognition and Cohesion Species

A mate-recognition species is a group of sexually reproducing organisms that recognize one another as potential mates. Expanding on this to allow for post-mating isolation, a cohesion species is the most inclusive population of individuals having the potential for phenotypic cohesion through intrinsic cohesion mechanisms; no matter whether populations can hybridize successfully, they are still distinct cohesion species if the amount of hybridization is insufficient to completely mix their respective gene pools. A further development of the recognition concept is provided by the biosemiotic concept of species.

Genetic Similarity and Barcode Species

In microbiology, genes can move freely even between distantly related bacteria, possibly extending to the whole bacterial domain. As a rule of thumb, microbiologists have assumed that kinds of Bacteria or Archaea with 16S ribosomal RNA gene sequences more similar than 97% to each other need to be checked by DNA-DNA hybridisation to decide if they belong to the same species or not. This concept was narrowed in 2006 to a similarity of 98.7%.

DNA-DNA hybridisation is outdated, and results have sometimes led to misleading conclusions about species, as with the pomarine and great skua. Modern approaches compare sequence similarity using computational methods.

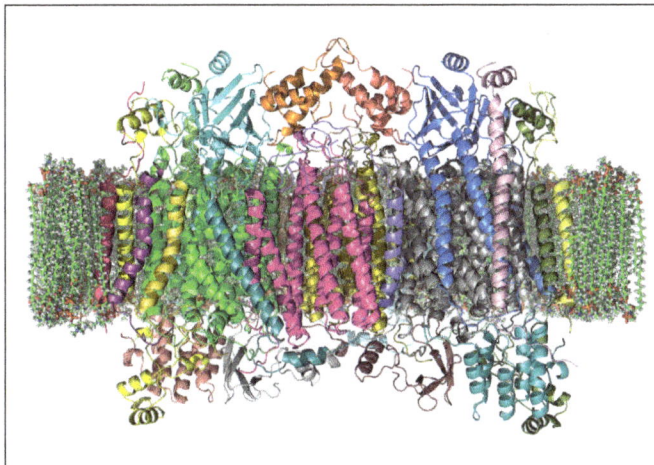

A region of the gene for the cytochrome c oxidase enzyme is used to distinguish speciesin the Barcode of Life Data Systems database.

DNA barcoding has been proposed as a way to distinguish species suitable even for non-specialists to use. The so-called barcode is a region of mitochondrial DNA within the gene for cytochrome c oxidase. A database, Barcode of Life Data Systems (BOLD) contains DNA barcode sequences from over 190,000 species. However, scientists such as Rob DeSalle have expressed concern that classical taxonomy and DNA barcoding, which they consider a misnomer, need to be reconciled, as they delimit species differently. Genetic introgression mediated by endosymbionts and other vectors can further make barcodes ineffective in the identification of species.

Phylogenetic, Cladistic or Evolutionary Species

A phylogenetic or cladistic species is an evolutionarily divergent lineage, one that has maintained its hereditary integrity through time and space. A cladistic species is the smallest group of populations that can be distinguished by a unique set of morphological or genetic traits. Molecular markers may be used to determine genetic similarities in the nuclear or mitochondrial DNA of various species. For example, in a study done on fungi, studying the nucleotide characters using cladistic species produced the most accurate results in recognising the numerous fungi species of all the concepts studied. Versions of the Phylogenetic Species Concept may emphasize monophyly or diagnosability.

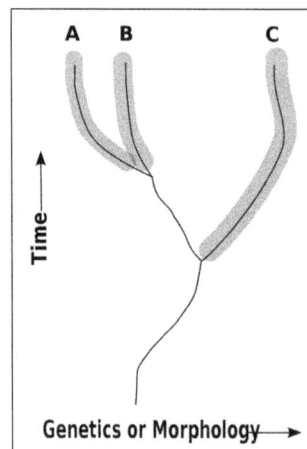

The cladistic or phylogenetic species concept is that a species is the smallest lineage which is distinguished by a unique set of either genetic or morphological traits.

Unlike the Biological Species Concept, a cladistic species does not rely on reproductive isolation, so it is independent of processes that are integral in other concepts. It works for asexual lineages, and can detect recent divergences, which the Morphological Species Concept cannot. However, it does not work in every situation, and may require more than one polymorphic locus to give an accurate result. The concept may lead to splitting of existing species, for example of Bovidae, into many new ones.

An evolutionary species, suggested by George Gaylord Simpson in 1951, is "an entity composed of organisms which maintains its identity from other such entities through time and over space, and which has its own independent evolutionary fate and historical tendencies". This differs from the biological species concept in embodying persistence over time. Wiley and Mayden state that they see the evolutionary species concept as "identical" to Willi Hennig's species-as-lineages concept, and assert that the biological species concept, "the several versions" of the phylogenetic species concept, and the idea that species are of the same kind as higher taxa are not suitable for biodiversity studies (with the intention of estimating the number of species accurately). They further suggest that the concept works for both asexual and sexually-reproducing species.

Ecological Species

An ecological species is a set of organisms adapted to a particular set of resources, called a niche, in the environment. According to this concept, populations form the discrete phenetic clusters that we recognise as species because the ecological and evolutionary processes controlling how resources are divided up tend to produce those clusters.

Genetic Species

A genetic species as defined by Robert Baker and Robert Bradley is a set of genetically isolated interbreeding populations. This is similar to Mayr's Biological Species Concept, but stresses genetic rather than reproductive isolation. In the 21st century, a genetic species can be established by comparing DNA sequences, but other methods were available earlier, such as comparing karyotypes (sets of chromosomes) and allozymes (enzyme variants).

Evolutionarily Significant Unit

An evolutionarily significant unit (ESU) or "wildlife species" is a population of organisms considered distinct for purposes of conservation.

Chronospecies

In palaeontology, with only comparative anatomy (morphology) from fossils as evidence, the concept of a chronospecies can be applied. During anagenesis (evolution, not necessarily involving branching), palaeontologists seek to identify a sequence of species, each one derived from the phyletically extinct one before through continuous, slow and more or less uniform change. In such a time sequence, palaeontologists assess how much change is required for a morphologically distinct form to be considered a different species from its ancestors.

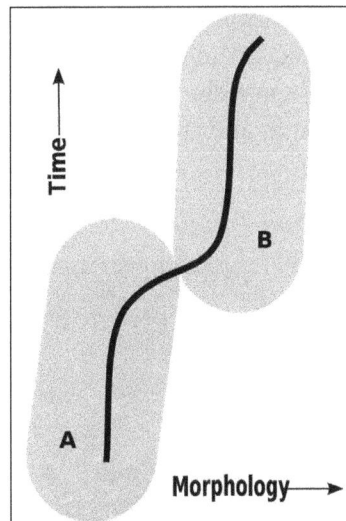

A chronospecies is defined in a single lineage (solid line) whose morphology changes with time. At some point, palaeontologists judge that enough change has occurred that two species (A and B), separated in time and anatomy, once existed.

Viral Quasispecies

Viruses have enormous populations, are doubtfully living since they consist of little more than a string of DNA or RNA in a protein coat, and mutate rapidly. All of these factors make conventional species concepts largely inapplicable. A viral quasispecies is a group of genotypes related by similar mutations, competing within a highly mutagenic environment, and hence governed by a mutation–selection balance. It is predicted that a viral quasispecies at a low but evolutionarily neutral and highly connected (that is, flat) region in the fitness landscape will outcompete a quasispecies located at a higher but narrower fitness peak in which the surrounding mutants are unfit, "the quasispecies effect" or the "survival of the flattest". There is no suggestion that a viral quasispecies resembles a traditional biological species.

Taxonomy and Naming

A cougar, mountain lion, panther, or puma, among other
common names: its scientific name is Puma concolor.

Common and Scientific Names

The commonly used names for kinds of organisms are often ambiguous: "cat" could mean the domestic cat, Felis catus, or the cat family, Felidae. Another problem with common names is that they often vary from place to place, so that puma, cougar, catamount, panther, painter and mountain lion all mean Puma concolor in various parts of America, while "panther" may also mean the jaguar (Panthera onca) of Latin America or the leopard (Panthera pardus) of Africa and Asia. In contrast, the scientific names of species are chosen to be unique and universal; they are in two parts used together: the genus as in Puma, and the specific epithet as in concolor.

Species Description

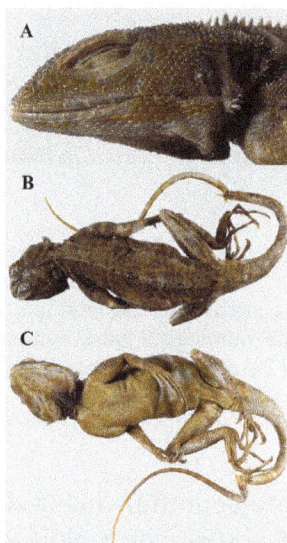

The type specimen (holotype) of Lacerta
plica, described by Linnaeus.

A species is given a taxonomic name when a type specimen is described formally, in a publication that assigns it a unique scientific name. The description typically provides means for identifying the new species, differentiating it from other previously described and related or confusable species and provides a validly published name (in botany) or an available name (in zoology) when the paper is accepted for publication. The type material is usually held in a permanent repository, often the research collection of a major museum or university, that allows independent verification and the means to compare specimens. Describers of new species are asked to choose names that, in the words of the International Code of Zoological Nomenclature, are "appropriate, compact, euphonious, memorable, and do not cause offence".

Abbreviations

Books and articles sometimes intentionally do not identify species fully and use the abbreviation "sp." in the singular or "spp." (standing for species pluralis, the Latin for multiple species) in the plural in place of the specific name or epithet (e.g. Canis sp.). This commonly occurs when authors are confident that some individuals belong to a particular genus but are not sure to which exact species they belong, as is common in paleontology. Authors may also use "spp." as a short way of saying that something applies to many species within a genus, but not to all. If scientists mean that something applies to all species within a genus, they use the genus name without the specific name or epithet. The names of genera and species are usually printed in italics. Abbreviations such as "sp." should not be italicised. When a species identity is not clear a specialist may use "cf." before the epithet to indicate that confirmation is required. The abbreviations "nr." (near) or "aff." (affine) may be used when the identity is unclear but when the species appears to be similar to the species mentioned after.

Identification Codes

With the rise of online databases, codes have been devised to provide identifiers for species that are already defined, including:

- National Center for Biotechnology Information (NCBI) employs a numeric 'taxid' or Taxonomy identifier, a "stable unique identifier", e.g., the taxid of Homo sapiens is 9606.

- Kyoto Encyclopedia of Genes and Genomes (KEGG) employs a three- or four-letter code for a limited number of organisms; in this code, for example, H. sapiens is simply hsa.

- UniProt employs an "organism mnemonic" of not more than five alphanumeric characters, e.g., HUMAN for H. sapiens.

- Integrated Taxonomic Information System (ITIS) provides a unique number for each species. The LSID for Homo sapiens is urn:lsid:catalogueoflife.org:taxon:4da6736d-d35f-11e6-9d3f-bc764e092680:col20170225.

Lumping and Splitting

The naming of a particular species, including which genus (and higher taxa) it is placed in, is a hypothesis about the evolutionary relationships and distinguishability of that group of organisms. As further information comes to hand, the hypothesis may be confirmed or refuted. Sometimes, especially in the past when communication was more difficult, taxonomists working in isolation have given two distinct names to individual organisms later identified as the same species. When two named species are discovered to be of the same species, the older species name is given priority and usually retained, and the newer name considered as a junior synonym, a process called synonymisation. Dividing a taxon into multiple, often new, taxa is called splitting. Taxonomists are often referred to as "lumpers" or "splitters" by their colleagues, depending on their personal approach to recognising differences or commonalities between organisms.

Broad and Narrow Senses

The nomenclatural codes that guide the naming of species, including the ICZN for animals and the ICN for plants, do not make rules for defining the boundaries of the species. Research can change the boundaries, also known as circumscription, based on new evidence. Species may then need to be distinguished by the boundary definitions used, and in such cases the names may be qualified with sensu stricto ("in the narrow sense") to denote usage in the exact meaning given by an author such as the person who named the species, while the antonym sensu lato ("in the broad sense") denotes a wider usage, for instance including other subspecies. Other abbreviations such as "auct." ("author"), and qualifiers such as "non" ("not") may be used to further clarify the sense in which the specified authors delineated or described the species.

Mayr's Biological Species Concept

Most modern textbooks make use of Ernst Mayr's 1942 definition, known as the Biological Species Concept as a basis for further discussion on the definition of species. It is also called a reproductive or isolation concept. This defines a species as

groups of actually or potentially interbreeding natural populations, which are reproductively isolated from other such groups.

It has been argued that this definition is a natural consequence of the effect of sexual reproduction on the dynamics of natural selection. Mayr's use of the adjective "potentially" has been a point of debate; some interpretations exclude unusual or artificial matings that occur only in captivity, or that involve animals capable of mating but that do not normally do so in the wild.

The Species Problem

It is difficult to define a species in a way that applies to all organisms. The debate about species delimitation is called the species problem. The problem was recognized even in 1859, when Darwin wrote in On the Origin of Species:

No one definition has satisfied all naturalists; yet every naturalist knows vaguely what he means when he speaks of a species. Generally the term includes the unknown element of a distinct act of creation.

When Mayr's concept breaks down.

Palaeontologists are limited to morphological evidence when deciding whether fossil life-forms like theseInoceramus bivalves formed a separate species.

A simple textbook definition, following Mayr's concept, works well for most multi-celled organisms, but breaks down in several situations:

- When organisms reproduce asexually, as in single-celled organisms such as bacteria and other prokaryotes, and parthenogenetic or apomictic multi-celled organisms. The term quasispecies is sometimes used for rapidly mutating entities like viruses.

- When scientists do not know whether two morphologically similar groups of organisms are capable of interbreeding; this is the case with all extinct life-forms in palaeontology, as breeding experiments are not possible.

- When hybridisation permits substantial gene flow between species.

- In ring species, when members of adjacent populations in a widely continuous distribution range interbreed successfully but members of more distant populations do not.

Species identification is made difficult by discordance between molecular and morphological investigations; these can be categorized as two types: (i) one morphology, multiple lineages (e.g. morphological convergence, cryptic species) and (ii) one lineage, multiple morphologies (e.g. phenotypic plasticity, multiple life-cycle stages). In addition, horizontal gene transfer (HGT) makes it difficult to define a species. All species definitions assume that an organism acquires its genes from one or two parents very like the "daughter" organism, but that is not what happens in HGT. There is strong evidence of HGT between very dissimilar groups of prokaryotes, and at least occasionally between dissimilar groups of eukaryotes, including some crustaceans and echinoderms.

The evolutionary biologist James Mallet concludes that there is no easy way to tell whether related geographic or temporal forms belong to the same or different species. Species gaps can be verified only locally and at a point of time. One is forced to admit that Darwin's insight is correct: any local reality or integrity of species is greatly reduced over large geographic ranges and time periods.

The willow warbler and chiffchaff are almost identical in appearance but do not interbreed.

Aggregates of Microspecies

The species concept is further weakened by the existence of microspecies, groups of organisms, including many plants, with very little genetic variability, usually forming species aggregates. For example, the dandelion Taraxacum officinale and the blackberry Rubus fruticosus are aggregates with many microspecies—perhaps 400 in the case of the blackberry and over 200 in the dandelion, complicated by hybridisation, apomixis and polyploidy, making gene flow between populations difficult to determine, and their taxonomy debatable. Species complexes occur in insects such as Heliconius butterflies, vertebrates such as Hypsiboas treefrogs, and fungi such as the fly agaric.

Aggregates of Microspecies

The species concept is further weakened by the existence of microspecies, groups of organisms, including many plants, with very little genetic variability, usually forming species aggregates. For example, the dandelion Taraxacum officinale and the blackberry Rubus fruticosus are aggregates with many microspecies—perhaps 400 in the case of the blackberry and over 200 in the dandelion, complicated by hybridisation, apomixis and polyploidy, making gene flow between populations difficult to determine, and their taxonomy debatable. Species complexes occur in insects such as Heliconius butterflies, vertebrates such as Hypsiboas treefrogs, and fungi such as the fly agaric.

Blackberries belong to any of hundreds of microspecies of the Rubus fruticosus species aggregate.

The butterfly genus Heliconius contains many similar species.

The Hypsiboas calcaratus–fasciatus species complex contains at least six species of treefrog.

Hybridisation

Natural hybridisation presents a challenge to the concept of a reproductively isolated species, as fertile hybrids permit gene flow between two populations. For example, the carrion crow Corvus corone and the hooded crow Corvus cornix appear and are classified as separate species, yet they hybridise freely where their geographical ranges overlap.

Hybridisation of carrion and hooded crows permits gene flow between 'species'

Carrion crow.

Hybrid with dark belly, dark gray nape.

Hybrid with dark belly.

Hooded crow.

Ring Species

A ring species is a connected series of neighbouring populations, each of which can sexually inter-breed with adjacent related populations, but for which there exist at least two "end" populations in the series, which are too distantly related to interbreed, though there is a potential gene flow between each "linked" population. Such non-breeding, though genetically connected, "end" populations may co-exist in the same region thus closing the ring. Ring species thus present a difficulty for any species concept that relies on reproductive isolation. However, ring species are at best rare. Proposed examples include the herring gull-lesser black-backed gull complex around the North pole, the Ensatina eschscholtzii group of 19 populations of salamanders in America, and the greenish warbler in Asia, but many so-called ring species have turned out to be the result of mis-classification leading to questions on whether there really are any ring species.

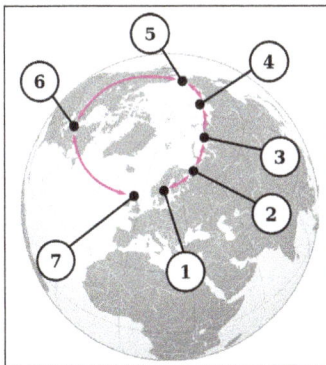

Seven "species" of Larus gulls interbreed in a ring around the Arctic.

Opposite ends of the ring: a herring gull (Larus argentatus) (front) and a lesser black-backed gull (Larus fuscus).

A greenish warbler, Phylloscopus trochiloides.

Presumed evolution of five "species" of greenish warblers around Himalayas.

Change

Species are subject to change, whether by evolving into new species, exchanging genes with other species, merging with other species or by becoming extinct.

Speciation

The evolutionary process by which biological populations evolve to become distinct or reproductively isolated as species is called speciation. Charles Darwin was the first to describe the role of natural selection in speciation in his 1859 book The Origin of Species. Speciation depends on a measure of reproductive isolation, a reduced gene flow. This occurs most easily in allopatric speciation, where populations are separated geographically and can diverge gradually as mutations accumulate. Reproductive isolation is threatened by hybridisation, but this can be selected against once a pair of populations have incompatible alleles of the same gene, as described in the Bateson–Dobzhansky–Muller model. A different mechanism, phyletic speciation, involves one lineage gradually changing over time into a new and distinct form, without increasing the number of resultant species.

Exchange of Genes between Species

Horizontal gene transfer between organisms of different species, either through hybridisation, antigenic shift, or reassortment, is sometimes an important source of genetic variation. Viruses can transfer genes between species. Bacteria can exchange plasmids with bacteria of other species, including some apparently distantly related ones in different phylogenetic domains, making analysis of their relationships difficult, and weakening the concept of a bacterial species.

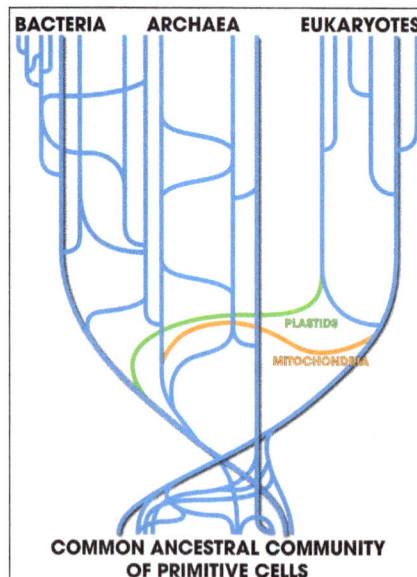

Horizontal gene transfers between widely separated
species complicate the phylogeny of bacteria.

Louis-Marie Bobay and Howard Ochman suggest, based on analysis of the genomes of many types of bacteria, that they can often be grouped "into communities that regularly swap genes", in much the same way that plants and animals can be grouped into reproductively isolated breeding populations. Bacteria may thus form species, analogous to Mayr's biological species concept, consisting of asexually reproducing populations that exchange genes by homologous recombination.

Extinction

A species is extinct when the last individual of that species dies, but it may be functionally extinct well before that moment. It is estimated that over 99 percent of all species that ever lived on Earth, some five billion species, are now extinct. Some of these were in mass extinctions such as those at the ends of the Permian, Triassic and Cretaceous periods. Mass extinctions had a variety of causes including volcanic activity, climate change, and changes in oceanic and atmospheric chemistry, and they in turn had major effects on Earth's ecology, atmosphere, land surface, and waters. Another form of extinction is through the assimilation of one species by another through hybridization. The resulting single species has been termed as a "compilospecies".

Practical Implications

Biologists and conservationists need to categorise and identify organisms in the course of their work. Difficulty assigning organisms reliably to a species constitutes a threat to the validity of research results, for example making measurements of how abundant a species is in an ecosystem moot. Surveys using a phylogenetic species concept reported 48% more species and accordingly smaller populations and ranges than those using nonphylogenetic concepts; this was termed "taxonomic inflation", which could cause a false appearance of change to the number of endangered species and consequent political and practical difficulties. Some observers claim that there is an inherent conflict between the desire to understand the processes of speciation and the need to identify and to categorise.

Conservation laws in many countries make special provisions to prevent species from going extinct. Hybridization zones between two species, one that is protected and one that is not, have sometimes led to conflicts between lawmakers, land owners and conservationists. One of the classic cases in North America is that of the protected northern spotted owl which hybridizes with the unprotected California spotted owl and the barred owl; this has led to legal debates. It has been argued that the species problem is created by the varied uses of the concept of species, and that the solution is to abandon it and all other taxonomic ranks, and use unranked monophyletic groups instead. It has been argued, too, that since species are not comparable, counting them is not a valid measure of biodiversity; alternative measures of phylogenetic biodiversity have been proposed.

References

- Barnes, richard stephen kent (2001). The invertebrates: a synthesis. Wiley-blackwell. P. 41. Isbn 978-0-632-04761-1

- Five-kingdoms-classification, biology: byjus.Com, retrieved 1 may, 2019

- Simpson, alastair g.B.; Roger, andrew j. (2004). "The real 'kingdoms' of eukaryotes". Current biology. 14 (17): R693–r696. Doi:10.1016/J.Cub.2004.08.038. Pmid 15341755

- Margulis l, chapman mj (2009-03-19). Kingdoms and domains: an illustrated guide to the phyla of life on earth. Academic press. Isbn 9780080920146

- Desjardin de, perry ba, lodge dj, stevani cv, nagasawa e (2010). "Luminescent mycena: new and noteworthy species". Mycologia. 102(2): 459–77. Doi:10.3852/09-197. Pmid 20361513

- "Food and drink". Kew gardens. Archived from the original on 28 march 2014. Retrieved 1 october 2017

- Berg, linda r. (2 March 2007). Introductory botany: plants, people, and the environment (2 ed.). Cengage learning. P. 15. Isbn 9780534466695. Retrieved 23 july 2012

- "Species concepts". Scientific american. 20 April 2012. Archived from the original on 14 march 2017. Retrieved 14 march2017

- Margulis, lynn; chapman, michael j. (2009). Kingdoms and domains (4th corrected ed.). London: academic press. Isbn 9780123736215

INDEX